QUEER MEXICO

QUEER MEXICO
Cinema and Television since 2000
Paul Julian Smith

Wayne State University Press
Detroit

Series Editor
David Gerstner
A complete listing of books in this series can be found online at www.wsupress.wayne.edu

© 2017 by Wayne State University Press, Detroit, Michigan 48201. All rights reserved. No part of this book may be reproduced without formal permission. Manufactured in the United States of America.

ISBN 978-0-8143-4274-9 (paperback); ISBN 978-0-8143-4275-6 (ebook)
Library of Congress Cataloging Number: 2017947569

Wayne State University Press
Leonard N. Simons Building
4809 Woodward Avenue
Detroit, Michigan 48201-1309

Visit us online at wsupress.wayne.edu

CONTENTS

Acknowledgments — vii

Introduction: Diversity with Style — 1

1. Festivals, Webseries, Porn — 7
Festival Trajectories: Mix — 10
Web Romance: *Al final del arcoíris* — 18
Porn Stories: *Corrupción mexicana* — 25
Loving the Real — 30

2. The Art Cinema of Julián Hernández — 33
From the Social to the Aesthetic: *El cielo dividido* — 40
From Film to Life: *Yo soy la felicidad de este mundo* — 46
In Praise of the Fragment: Three Shorts — 52
The Wagers of Art — 57

3. Transgender Documentary — 61
Social Commitments: *Morir de pie* — 66
Media Memories: *Quebranto* — 70
Songs of Seduction: *Made in Bangkok* — 77
Lives in the Mirror — 82

4. Mainstream Movies — 85
Lesbian Romance: *Todo el mundo tiene a alguien menos yo* — 89
Ages of Gay Man: *Cuatro lunas* — 94
Trans Thriller: *Carmín tropical* — 101
A Gift of Love — 104

5. Argos's Queer Telenovela — 107
Family Plots: *La vida en el espejo* — 113
The Dereliction of Masculinity: *El sexo débil* — 120
Women behind Bars: *Capadocia* — 128
Intimate Strangers — 133

Conclusion: Two Films, Two Futures — 137

Appendix: Interview with Julián Hernández at the Morelia International Film Festival, 2014   143

Works Cited   151

Index   161

ACKNOWLEDGMENTS

I thank above all my students and colleagues at the Graduate Center of the City University of New York, especially in the Hispanic and Luso-Brazilian Program, which has proved such a fruitful place to carry out my research. Lily Ryan, a doctoral candidate at the GC, was the research assistant who compiled the impeccable references and bibliography. Rojo Robles kindly recommended bibliography on cinephilia.

My thanks go also to the two anonymous readers of the manuscript for their helpful comments and to Annie Martin at Wayne State University Press for her kindness and efficiency. I am most grateful also to B. Ruby Rich, a queer icon at the Guadalajara International Film Festival and editor of *Film Quarterly*, who has given me, as columnist, an opportunity to write on film and television of the Spanish-speaking world. Some of the material included here in chapters 3 and 4 was published in *FQ* in early, short versions, as was the interview in the appendix.

Raúl Miranda, the head of the Centro de Documentación at the Cineteca in Mexico City, has been consistently kind and helpful. He introduced me to two of the features I study in chapter 3. The annual Coloquios Internacionales Cine Iberoamericano Contemporáneo, which I have co-organized with Nancy Berthier of the Sorbonne, Antonia del Rey Reguillo of Valencia, and Álvaro A. Fernández of Guadalajara, have been invaluable. I am also grateful for invitations to the festivals of Guadalajara and Morelia where I learned so much about media industries in Mexico. Guillermo Orozco has been an inspiration in Mexican TV studies, as has Carlos Bonfil in Mexican film criticism. Fernanda Solórzano, another essential film critic, has been consistently supportive. Sue Dugen (MSocInd) compiled the excellent index.

My greatest debts, however, are to Julián Hernández, who generously provided the cover image and sat for the interview included here as an appendix, and to David Gerstner, who so kindly commissioned this book as the first in the collection that he is editing for Wayne State and has so warmly encouraged me throughout this process.

This book is dedicated to all of the creators whose work is analyzed and celebrated in it.

INTRODUCTION
Diversity with Style

This book is the first to treat LGBT media in the Mexico of the twenty-first century. Paying close attention to Mexican cultural practice, it combines textual analysis of audiovisual works as aesthetic objects with an institutional account of the conditions of production, distribution, and exhibition of those works across the vibrant cinema and television sectors in a rapidly changing modern Mexico. Crucially it extends the focus of attention beyond fiction feature films and art cinema to embrace less familiar media such as *webnovela*, pornography, shorts, documentary, and television series. *Queer Mexico* thus aims to tease out the relationships between art and commerce, identity and aesthetics, and the artistic and social turns of queer Mexican media that are irreducibly hybrid.

This introduction will start to sketch a social background to LGBT representation in Mexican media, citing materials that range from essays by renowned cultural commentator Carlos Monsiváis to histories of regional LGBT activism and academic attempts to rehabilitate the distinctively Mexican slur *joto* ("fairy"). It will also mention briefly the oeuvre of a rare but influential auteur of the previous century, Jaime Humberto Hermosillo.

Queer Mexico proper is divided into five chapters. The first treats three discursive contexts for the audiovisual fiction texts that follow: a long-lasting LGBT festival, a web-distributed TV youth drama, claimed by its makers to be the first wholly gay series made in Mexico, and an equally well established local porn producer, rare indeed in Latin America. All of these three areas refer back to a heritage of queer imagery in Mexico (the porn videos feature *lucha libre* wrestlers), even as they point forward to new cultural identities. And all are the creation of little-known minor entrepreneurs who have cultivated their respective and neglected fields over time.

The second chapter examines selected features and shorts by Mexico's sole internationally distributed gay art house director, Julián Hernández. Relatively marginalized in Mexico, Hernández represents nonetheless a radical challenge to the canonic heterosexual auteurs who are generally taken to represent his country abroad, even though the art movies by those directors, feted at foreign festivals, are little seen at home. Hernández's practice of auteur cinema is juxtaposed in this book with two parallel narratives: the real-life micronarratives of current queer lives narrated by a social scientist and the homosocial "cinemachismo" examined by a scholar of Mexico's Golden Age cinema of the 1940s. At the time of writing (April 2016), shorts by Hernández and his producer Roberto Fiesco, once difficult to access, are available in the United States from Amazon Instant Video under the teasing title *Mexican Men* (2016). As an appendix I include here a lengthy interview with the articulate Hernández, which he kindly granted me at the Morelia International Film Festival.

The third chapter of the book treats the rising genre of documentary on transgender themes. While it is often said that Mexican documentary is more creative than its fiction rival (and the influential Morelia festival has long been known for its fostering of the genre), it is only recently that queer documentaries have made their presence felt. Strikingly, like the "minor genres" treated in the previous chapter, several of these documentary features cite Mexican audiovisual heritage. They thus work through inherited image repertoires in cinema and politics (from child stars of the Golden Age to Che Guevara), even as they push the boundaries of what counts as Mexican national cinema, moving beyond the nation's geographical borders to Cuba and even Thailand. This question of location or physical space (which often goes unrecognized by foreign audiences) will prove central to most of my chapters.

The fourth chapter charts the growing trend of a gay male-, lesbian-, or trans-focused mainstream cinema, a phenomenon treated outside Mexico by Juett and Jones's *Coming Out to the Mainstream* (2010) and Rich's *New Queer Cinema* (2013). Such films reject both the gritty social realism and the tragic melodrama of earlier periods, drawing rather on the massively popular genres of the romantic comedy and the thriller that attract aspirational audiences to theaters. These movies thus offer their target public reassuringly positive images of socially integrated queers. The relatively assimilated characters depicted in this commercial cinema (which include lesbian and gay professionals and a self-confident transgender woman) are also very different from the alienated and isolated protagonists seen in the auteur films of the second chapter.

The final chapter, the longest, examines the rich and diverse history of queer representation in Mexico's dominant TV genre and, arguably, national narrative: the telenovela. It focuses especially on Argos, a politically progressive independent production company that has made gay- and lesbian-friendly serial melodramas for some twenty years. Surprisingly well received and widely distributed, these telenovelas have played to audiences in their millions, with Argos collaborating over time across a range of television channels in Mexico: from free to air network Azteca, to radical upstart Cadena 3, via boundary-pushing subscription channel HBO Latin America. Despised by the elite in Mexico, television serves nonetheless the vital purpose of inviting the "intimate strangers" that are queer characters into family homes, an opportunity that cinema, of course, is unable to offer.

As a whole the book thus hopes to demonstrate the diversity of both representations and production processes in this corpus of audiovisual works. It attempts also to reconstruct a queer cultural field for Mexico that incorporates multiple texts, producers, and institutions: from auteur cinema to melodrama and from film festivals to porn studios. It is a queer field that makes itself felt, necessarily, within the broader, allegedly unmarked field of mainstream Mexican cultural production, which alternately marginalizes and promotes it. One recurring theme from industry professionals such as Hernández and Fiesco, however, is that in Mexican cinema, and perhaps television also, open homophobia is no longer socially acceptable.

Anecdotally, young queers in the capital also seem confidently visible, happy to express affection in public. Mexico City's iconic Latin American Tower, a glass and steel calque of the Empire State Building, was once a location for heterosexual romantic comedies such as Alfonso Cuarón's *Solo con tu pareja* ("Only with Your Partner," 1991). On my recent visit to its viewing platform, where the megalopolis can be seen stretching out forever toward the volcanoes of the Valley of Mexico, no one turned a hair as a young lesbian couple shared a passionate kiss at sunset. Outside also male couples were smooching in the gathering dusk by the tumbledown eighteenth-century San Francisco church in the shadow of the Tower.

It seems likely, then, that in certain *colonias* of the capital at least (the Historic Center and the gay village of the Zona Rosa) and in spite of continuing homo- and trans-phobia elsewhere, a new Mexico exists for queer people that belies macho or conservative Catholic stereotypes. Public visibility is matched by marriage equality, a reality, in Mexico City again (but not elsewhere), since 2010. And in 2013 the Mexican Supreme Court ruled that the use of homophobic slurs was a violation of fundamental human rights.

A flurry of books has also appeared. One is a posthumous collection of essays by respected intellectual Carlos Monsiváis (2010) on "sexual diversity" (the preferred term in Mexico), which focuses on issues such as the continuing coexistence of a traditional gendered model of queerness (male/female identification) with the newer international model of object choice (gay/straight affiliation). Monsiváis's queer chronicle of Mexico begins with the sadly famous "Dance of the 41," where police raided a Mexico City transvestite ball in 1901 (2010, 82–84), a symbolic beginning for a visibly queer culture. Another book offers a history of LGBT activism in the conservative northern metropolis of Monterrey, where the film festival offers a rare oasis of tolerance (Quintero Murguía 2015). And a third attempts an extended academic rehabilitation of the slur *joto* in the cultural field, from literature to theater (Marquet 2006). It is a labor carried out in a more popular register by a fourth "chronology and dictionary" of *jotos* (Cobian 2013). All are available for purchase at Voces en Tinta (Voices in Ink), an established queer book store, café, and community center in Mexico City's Zona Rosa.

Lifestyle magazine *Divers* (motto "Diversity with Style") also offers a glimpse into this newly confident culture of queer visibility and sociability. I bought the issue for December 2015 in a branch of Sanborns, the ubiquitous restaurant-cum-store franchise owned by Carlos Slim, Mexico's richest man. Stuffed as it is with luxury advertising, the magazine reveals a transparent consumerism. It boasts travel articles on Zurich, Tenerife, and "New York in winter," destinations surely inaccessible to the overwhelming majority of *chilangos* (Mexico City residents). Yet the issue is not devoid of activism. There is one article on the late Nancy Cárdenas, who founded Mexico City's Homosexual Liberation Front as early as 1971, and another on the "Blue Ribbon Boys" who currently promote sexual health and HIV awareness via an app. And in a third the local mayor (officially "Head of Government" of an urban behemoth twenty-six million strong) poses proudly at a ceremony where the Mexican capital joined such enlightened places as San Francisco, Berlin, and Buenos Aires as an officially "LGBTTTI-friendly" city.

Such changes did not take place in a vacuum. In 2000, the date at which this book begins, there was a historic change in political regime, when the PRI (or Institutional Revolutionary Party) was finally voted out of national office after some seventy years. Although Mexicans constantly complain about democratic deficits in their country, and with good reason, there is now for the first time a relative openness to criticism of the state and no direct censorship of culture. This process has also been much studied in different areas. In the political sphere Gavin O'Toole's *The Reinvention of Mexico* (2010) examines the decline of the official PRI ideology of revolutionary (and cultural) nationalism, as the country opened up its nationalized industries to international markets, especially the United States. In the field of film the challenge posed by "neoliberalism" is linked by Ignacio Sánchez Prado (2014) to such varied phenomena as the romantic comedies of the 1990s, like Cuarón's *Solo con tu pareja*, and the uncompromising art films of Carlos Reygadas in the 2000s.

What is clear, however, within Mexico, yet often invisible to observers outside, is the recent rise of a new and large middle class, which has meant that the country is, according to economic criteria, no longer a developing but rather a middle-income country. Of course, Mexico remains riven by inequality. The *Economist*'s special supplement "The Two Mexicos" (September 19–25, 2015) cites a continuing chasm between the industrially developed north and capital and the underdeveloped southern states, racked by violence and institutionalized corruption (Mexico City is, on the other hand, now safer than it was).

The increase in size and wealth of the middle class, however, which is very visible in many of the texts I examine in this book, is missed by first world festivalgoers mesmerized by the ultraviolence and squalor offered them by heterosexual auteurs such as Reygadas and his disciple Amat Escalante. This is not to deny that violence and poverty remain sadly topical themes in Mexico. But the fact that they monopolize foreign interest in the country, both academic and general, is highly disturbing. Moreover, the apocalyptic vision of Mexico promoted in the press, especially in the United States, is not borne out by the films and TV shows I treat here.

Art historian Cuauhtémoc Medina has added a further twist to accounts of changes in Mexican politics and culture in the millennium, arguing for a decline in the literary ethos that was once vital to national life and the consequent rise in the role of the visual sphere (Aguirre 2012). It is an emphasis that is seen also in academic work in the field, which is now more often addressing visual culture.

The admirable collection of essays *México se escribe con J* ("Mexico Is Spelled with a J," Schuessler and Capistrán 2010), a title that puns on the letter *J* (pronounced "jota") and *joto* meaning "queer," has in spite of its essentially literary focus contributions on cinema and even television, a medium generally ignored. In the field of serious Mexican journalism, special mention should be made of Carlos Bonfil in cinema and Álvaro Cueva in television, both of whom have treated queer issues in their respective media. Readers will perhaps be surprised to see that I also cite gossip magazines *TVyNovelas* and *TVNotas* for the evidence they reveal of fans' shifting attitudes to LGBT celebrities and their families, as they are drawn into the media spotlight, proudly or reluctantly as the case may be.

Yet in Mexican gay film studies in the United States, especially, there has been a concentration on just two films released before my time period. Arturo Ripstein's *El lugar sin límites* ("The Place without Limits," 1978) tells the tale of a rural transvestite's tragic affair with a Mexican macho. Jaime Humberto Hermosillo's comic *Doña Herlinda y su hijo* ("Doña Herlinda and Her Son," 1985), on the other hand, is the story of a matriarch in conservative Guadalajara who manages to incorporate her son's male lover into her respectable household. It is not self-evident how relevant these famous features are to current circumstances. Yet this continuing focus on a restricted canon is confirmed by two still-recent books. Aarón Díaz Mendiburo examines the "homoerotic *hijos*" ("children" or "sons") of Jaime Humberto Hermosillo (2004), offering close readings of five films by the respected and durable openly gay maestro, with no reference to his successors. Bernard Schulz-Cruz (2008) gives rather a survey of over

Introduction - 5

thirty films on the gay theme (including the two mentioned above) made during what he calls "three decades of queerness" (*joterío*), from 1970 to 1999.

Given this social, cultural, and scholarly background, it seems urgent to me to address the period since 2000, hitherto neglected by scholars, and to expand the object of study beyond art cinema and more broadly feature film to address audiovisual texts and cultures in the neglected but vast and influential media of shorts, television, and webnovela. This will also involve the expansion of the audience addressed, as young people and those outside the capital are much less likely to have access to (and, indeed, interest in) auteur films than more senior metropolitan viewers. *Queer Mexico* is thus intended as a gesture of respect to the producers and consumers of this newly diverse audiovisual culture. And I hope that readers of this book will share my excitement in discovering a recent body of work that transcends foreign stereotypes of both nationality and sexuality, treating diversity of desire with cinematic and televisual style.

FESTIVALS, WEBSERIES, PORN

The camera moves slowly over an attractive male couple on a city balcony, clad only in speedos. We hear them argue as one sings languidly along to Nancy Sinatra's "Bang Bang (My Baby Shot Me Down)." The other man, who has handsome mestizo features and a dancer's physique, leaves the apartment to sunbathe nude on their building's roof. An attractive young woman approaches him and initiates sex (during the act he flashes back to his male lover's face). When they have finished, the boy discovers that his clothes have been stolen. The girl then lends him her flimsy, flowery frock to return home in. His boyfriend, aroused at this unaccustomed costume, makes love to him on the kitchen counter. As they embrace, the partner flashes back to the girl he had just met on the roof.

This short video, which is made up entirely of still photos, is the trailer for the eighteenth edition of the Mix Festival (known more formally as "Festival of Sexual Diversity in Film and Video"), which was held in 2014 in Mexico City. And in this opening chapter I address three minor genres, practices, or discursive contexts that show better perhaps than major feature films how the audiovisual is embedded in everyday Mexican queer life. Once marginal, these practices are now central to academic inquiry (festivals are an established topic at, say, the annual conference of Society of Cinema and Media Studies). Yet my three objects of analysis remain little studied, indeed relatively little known, even in Mexico, their home country. They are the already mentioned Mix Festival; a teenage webseries made by small independent multimedia company Tres Tercios, said to be first wholly gay drama in the country; and a long-lasting porn producer called Mecos, allegedly named after an indigenous word for "cum."

Although these three phenomena are clearly diverse in their production and reception, I will argue that they have much in common. To anticipate, they all embrace hybridity, transmedia, and entrepreneurship. Thus, as we shall see, the festival stages tensions between the domestic and the foreign and between the cultural and the erotic. The webseries combines a romantic, even utopian, vision of queer youth "at the end of the rainbow" with a more everyday social perspective, which acknowledges the necessary limits of gay life in the metropolis, especially for high school students. And the porn videos, surprisingly perhaps, head in a similar direction, moving out from the utopian promise of the purely erotic single scene to embrace (however slyly or ironically) more troubling social issues that are expressed in narrative form and at feature length. In all three cases the combination of existing artistic elements can thus lead to innovation, to the emergence of new sexual and artistic forms.

Hybrid by nature, my three examples are also transmediatic in their creation and distribution. Marginal to established cultural channels, they leave of necessity just a fragile remainder in print or on the web. Thus, the festival lives on only in its programs, posters, and promos (I consulted materials for early editions on paper in the library or Center of Documentation in Mexico City's Cineteca). The webseries vanished from the Internet after its first staggered release and was only briefly available on DVD (I purchased it at a Mixup video store in the capital's Historic Center). Mecos's films are also edited on DVD (I bought one in a sex shop in the gay village of the Zona Rosa), but until recently are more likely to be accessed in blurry, roughly edited versions on free aggregation sites such as PornHub and XTube (Mecos's online store is only accessible within Mexico).

Yet perhaps, as deconstruction taught us long ago, the center is dependent on the margin that it seeks in vain to exclude, marking as the latter does the limit of acceptable expression. Certainly, in all three cases these unauthorized and often collective practices are also the creation of little known individual auteurs, who, like their more consecrated counterparts, have long labored to produce original cultural work. Indeed, there are close and sometimes surprising links between my minor genres and mainstream audiovisual practitioners such as art house director Julián Hernández, the subject of my next chapter.

Entrepreneurship is also more complex than might first appear. One trade journal claims that it requires attention not only to ideas and profits but also to social goods (Brooks 2015). The entrepreneur thus engages simultaneously with the registers of the creative, the financial, and even the ethical. And his or her most important prerequisite is said to be "passion." It follows, then, that in my context of minor entrepreneurship, investment in all these areas (festivals, webseries, porn production) is to be read as both commercial and libidinal. Indeed, given the uncertain economics of the queer audiovisual in Mexico, where no such business is likely to make much money, the pleasure of the producers (as well as that of the consumers) will doubtless take precedence over their profit.

The festival video with which I began this chapter bears the name *Vestido* or "Dressed" in honor of Mix's theme that year. And it embodies the three characteristics mentioned earlier. Thus, it is hybrid in form, at once sweetly romantic and cheekily erotic, combining pop culture (that campy Nancy Sinatra song) with higher art: the trailer as a whole is in fact closely based on a short by established festival favorite François Ozon called *Une Robe d'été* ("A Summer Dress," 1996). The Mexican remake's refusal of moving pictures also mimics the technique of *La Jetée* (1962) by avant-garde pioneer Chris Marker. Mix's boy and girl even signal the trailer's debt to French cinephile tradition by exchanging a few words in that language.

The promo is also transmediatic, an insubstantial remainder of an event that now lives on only in the memories of festivalgoers and organizers or in the press cuttings of specialist reviewers. Indeed, it barely survives as a film itself: the version on YouTube is lacking in the crucial soundtrack, removed for copyright violation, and only the artier platform Vimeo currently hosts the full film.

Yet this ephemeral short also testifies obliquely to a sustained effort of entrepreneurship: it was produced by the founder and artistic director of Mix who has worked indefatigably for the festival for almost twenty years. And his declared mission to subvert or challenge viewers is clear even here in this trailer. The theme of "sexual diversity" (the label by which the festival has been officially named throughout its lifetime) is in *Vestido* dramatized by a close (too close for comfort?) cohabitation between straight and gay, women and men. The boy's first time with a girl, shown in the trailer, thus intermingles with his continuing affair with his male partner. It is a morale that would not necessarily prove palatable to mainstream gay audiences. On the other hand, the festival poster for that year is more reassuringly homoerotic: it shows rather the two men embracing in the kitchen, one naked and the other skimpily dressed in the newly torn summer dress. The girl, who is played by well-known film and TV actress Paulina Gaitán, is nowhere to be seen (she also appeared in *Las Aparicio* [Cadena 3/Argos, 2010], a telenovela, and later a feature film, with a feminist and lesbian theme).

Finally, the three phenomena I discuss in this chapter share a structuring metaphor: the journey or, even, "odyssey." The latter is a word used by both the festival of its successive editions and the webseries of its developing characters. The porn producer, in addition to a series of "national selections" or model auditions located in its studios in the capital's trendy *colonia* of La Roma, also offers in the most ambitious of its features an expedition exploring "corruption" across Mexico. Tracing the movement of queer bodies in time and space, then, the journey is a motif aptly suited for the study of these unsung audiovisual practices that have helped to create LGBT spaces and audiences in Mexico over almost two decades since the millennium.

Festival Trajectories: Mix (1997–present)

Mix Quince Años: XV Festival de Diversidad Sexual en Cine y Video

Fijación visual . . .

A quince años de nuestro nacimiento del Festival Mix se ha consolidado ya como el escaparate formal y arriesgado de las formas de vivir la diversidad sexual en el mundo. Con la cámara como testigo hemos recorrido—sin fronteras—caminos de deseos, sueños y necesidades que hace quince años considerábamos imposibles. . . . El cine ha sido arma de denuncia, elemento indispensable para incendiar la intimidad, generar la discusión y la comprensión, así como un lienzo de platino para descubrir nuevos discursos estéticos, retratos ocultos o visiones de un futuro brillante. . . . Y ahora gracias a ustedes, motivo de celebración: gracias por acompañarnos estos primeros quince años. Nuestra presentación a la sociedad ya se ha dado; ahora sigamos buscando la inserción bajo nuestros propios términos en una convivencia pacífica donde podamos mezclar ideas con libertad e inteligencia—sin etiquetas de por medio que sirvan de pretexto para detener este proceso de crecimiento democrático. Fijación visual: sin quitar la vista de lo que queremos ver, mostrar, interrogar, dialogar. . . . Pantalla para todos! (Arturo Castelán, 2011)

Fifteen Years of Mix: 15th Festival of Sexual Diversity in Cinema and Video

Visual fixation . . .

Fifteen years after our birth the Mix Festival has now established itself as the official and risky shop window for ways of living sexual diversity in the world. With the camera as our witness we have traveled—without borders—the paths of desires, dreams, and needs that fifteen years ago we thought impossible. . . . Cinema has been a weapon for social denunciation, an indispensable element for striking up intimacy, generating discussion and understanding, as well as a silver screen for discovering new aesthetic discourses, hidden portraits, or visions of a brilliant future. . . . And now, thanks to you, it is a cause for celebration: thanks for accompanying us for these first fifteen years. We have already introduced ourselves to society; now let's keep seeking our integration on our own terms in a peaceful coexistence where we can mix up ideas with freedom and intelligence—without any labels in the way that serve as a pretext to stop this process of democratic growth. Visual fixation: let's keep our eyes on what we want to see, show, question, discuss. . . . A screen for everyone!

This opening text from the 2011 edition of Mix is written, as ever, by Arturo Castelán, the founder and artistic director of the festival since its inauguration in 1997. It stresses first the "journey" traveled by organizers and viewers over a period of time that has now culminated in the fifteenth anniversary, the traditional age of maturity for Latinas commemorated in the *quinceañera*. Mix thus journeyed with (and contributed to) a social change in Mexico in a period when, as director Julián Hernández told me in an interview included at the end of this book, overt homophobia became no longer socially acceptable, in the film community at least.

Beyond this laudable continuity, the text is packed with contradictions and hybridities. Thus, Mix is said to be at once "official" (the Spanish word used is "formal," also used of a steady boyfriend) and "risky." Yet this official festival's journey knows no borders and it arrives at a destination that was previously unimaginable.

Mix is moreover an instrument for public activism ("social denunciation") but also for the structuring of private intimacy and the forging of artistic novelty. The festival organizers ("we") address themselves directly to their long-term public (also "we"), proposing a social or political project for (unnamed) sexual dissidents: having introduced themselves to Mexican society, they still need to be integrated into it, but on their own terms. Such a process is consciously nonconfrontational, seeking peaceful coexistence with a majority that, however, must be able to recognize and name the minority that seeks to be integrated into it. But the festival is also founded on fluidity (a "lack of labels"). Paradoxically once more the theme of the festival this year is not instability but "visual fixation," a term used knowingly here in its psychic as well as social sense.

Two panels on festivals at the Society of Cinema and Media Studies meeting in 2015 presented similar conflicts or contradictions in a more general manner, thus revealing current critical debates in the field. One, whose respondent was Tamara Falikov, one of the most distinguished scholars of festivals in Latin America, dealt with "The Challenges of Curating Latin American Film in the 21st Century." Here festival directors themselves gave presentations addressing the new "hybridity" of what are now known as "festival films" (Diana Sánchez) and the question of whether Latino festivals constitute "An Illusory Sense of Belonging or the Connecting Bridge of Communities" (Diana Vargas).

This problem of nationality or community is further complicated, as so often, by queerness. A second panel on "Speaking in (Queer) Tongues: LGBT Film Festivals and the Politics of Language" explored the thorny question of labeling (which vexed Mix), with special reference to Canada, a nation that is, like Mexico, a polyglot state ambivalent about its English-speaking superpower neighbor. One paper here explored "(Queer) Festival Programming as Translation" (Antoine Damiens), while a second meditated on "On Representation, Language, and Sexuality" (Ger Zielinski), and a third asked more explicitly "'What's in a Name?' The Language of Labelling Queer Film Festivals" (Stuart Richards). It should be said that Mix, to its credit, had anticipated these debates, appealing from the start to "sexual diversity." It thus did not need to undergo the changes in title experienced by events that were previously held under the banner "lesbian and gay" or even just "gay."

A recent report in *Film Quarterly* on India's sole mainstream LGBT festival (called Kashish or "allurement") brings these themes into closer focus, suggesting surprising correspondences with Mexico once more. Ani Maitra (2015) documents a paradoxical event: a state-supported queer festival in a country where (unlike in Mexico) homosexuality is still a criminal offense (60). Maitra writes that "without apparent irony" Kashish combined (like Mix) "incitement to rebellion" with (once more like Mix) a desire to "universalize queerness." The Mumbai festival thus exhibits a "contradictory impulse to engage in dialogue with the body politic while maintaining a queer resistance to it." Moreover Kashish, writes Maitra, took care to incorporate "cultural differences within [its] country," examining the representation of *hijras* (traditional local transgender subjects analogous to what are called *muxes* in Mexico) (62).

Yet Kashish also addressed globalization (what Mix called "no borders"). Tensions were highest at a panel featuring Australian queer theorist Dennis Altman (Maitra 2015, 64). While Altman had written in 1997 (according to Maitra once more) that "Western LGBT identity politics has a global reach and can transcend regional differences," he now emphasized rather "the fluidity of sexuality, the instability of sexual desire that labels like 'lesbian,' 'gay,' and 'transgender' fail to capture."

Maitra does not take this second argument seriously, however, because it fails to address the status of identity politics in an Indian context where, confusingly, homosexuality has recently been recriminalized but affirmative action has been legislated with respect to *hijras* and other transgender subjects. Maitra thus critiques in the context of queer festivals both the universalism of Western sexual identity politics and the false particularity that proposes fluidity as a panacea without engaging with local formulations of law and desire.

Mix clearly invokes the universal also, calling finally in the presentation cited above for "one screen for all." Yet Mix's media image is individual, not collective. Founder-director Castelán is the face of the festival, appearing in countless press, TV, and web interviews and authoring all of the program introductions. To use a Spanish idiom, he bravely "da la cara" (literally "gives his face," metaphorically "faces the music"), often confronting criticism even from within his own queer constituency. Moreover, the complicity between the national and the international visible in Kashish is seen here in the circle of legitimation in which the festival director himself participates: Castelán imports much programming from abroad (indeed the first edition of Mix México was based on preexisting festivals of the same name in New York and Brazil), and he is in turn exported into foreign film circuits (he served on the jury of Berlin's prestigious LGBT Teddy award).

Despite such hybridities (fixity/fluidity, domestic/foreign), the poster celebrating Mix's *quinceañera* boasts an unambiguous image: a European-featured young man with light brown hair lolls naked on a bed, smiling seductively at the camera (in later editions the festival will post on Facebook teasing behind-the-scenes videos of its annual "Mix boy" shoot). And if we examine more closely the texts and images in Mix's programs over the years, the contradiction between rebellion and institutionalization (or between erotics and aesthetics) becomes more evident.

Thus, the second festival program in 1998 seeks to establish the fledging event's legitimacy (Mix 1998). Although Castelán's text opens with an evocation of subjective fluidity and cites an "interminable flux of values and perspectives with respect to sexuality and art," it soon stresses objective criteria of value in local, regional, and global contexts. Hence the event is the only one of its kind in "Hispano-America" (a formulation that excludes Brazil), is supported by Mexico's Fondo Nacional para la Cultura y las Artes (National Fund for Culture and the Arts, later to give way to CONACULTA, the national culture council), and collaborates with "sibling festivals" in Los Angeles and Freiburg, Germany. The selection policy is "rigorous," with all titles being Mexican premieres or reappraisals in a new queer context. Subject matters and styles are also said to be "absolutely avant-garde."

The third festival introduction begins in 1999 with a sketch of queer film history that has a marked US bias: from Edison's short of two men dancing via Stonewall and Judy Garland's funeral to AIDS activism and "lesbian chic" (the phrase is written in English) (Mix 1999). In further evidence of an English-speaking bias, the documentary on gays and lesbians in Hollywood based on Vito Russo's *The Celluloid Closet* (1995) had even been screened without Spanish subtitles in the festival's first edition. If, according to the introduction again, the only constant in the culture of desire is historical "change," Mix is embedded ever more closely in a global and local network: the siblings are now given as Mix New York and Brazil, while the Mexican "children" are the festival's new offshoots in the regional centers Guadalajara, León, and Monterrey. The fourth festival ("Century Mix" in 2000) will thank the Cineteca Nacional (National Film Institute) for offering the main screening venue and global brands Kodak and Diesel for sponsoring named prizes (Mix 2000).

By the next edition in 2001 (called, inevitably perhaps, "Odyssey"), the festival, now much grown, claims one hundred national premieres from Australia, Canada, Spain, France, the USA, UK, Portugal, and Mexico (in that order) (Mix 2001). And the venues have expanded to include the National Library, at that time housed in a handsome period building in the Historic Center. "Supernova" (Mix 2002) begins with another literary reference, an abstract evocation of passion and ecstasy ascribed to Virginia Woolf. But by now the imagery on the programs, previously abstract also, is becoming more overtly sexual. The cover for 2003 boasts a huge, blown-up pair of male lips; 2004 a lower abdomen, barely covered by a groping hand; 2005 the tightly framed face of a stubbled Adonis; 2006 a shirtless muscular male caressed by a bunch of calla lilies, the iconic flowers famously painted by Diego Rivera as sensual accessories to female figures who are sometimes shown nude.

Homoeroticism thus joins hands with literature, fine art, and film history. The festival's 2007 edition ("Persona") queers Bergman's classic drama with two young men replacing his actresses, one in profile, the other face on. The cover for 2009 has a youth reclining on an armchair in his underpants, but with a hefty book lying on the floor within easy reach (the opening film was Julián Hernández's challenging *Rabioso sol, rabioso cielo* ["Raging Sun, Raging Sky"]). The 2010 event, named for the centenary of the Mexican Revolution, invokes Frida Kahlo, replaying her famed double self-portrait with an image of two boys posing in a ruffled skirt and tight corset.

By the time we reach 2013 the audience address is frankly libidinal: Castelán claims that the now mature festival's taste for "provocation" is intact, as proved by this year's poster and trailer, now posted on the web (Mix 2013). Boasting "young stars of cinema and dance" and shot in the Mexico City bathhouse SoDoMe, the promo materials of "Mix Dominante" have a light S/M feel, with its eight stars gussied up in fetish lingerie. One girl (Paulina Gaitán, once more) playfully spanks another or leads a bearded jock-strapped man on a leash. A transgender woman poses next to a bare bottom. And four naked boys, of diverse ethnic varieties, are shown intertwined, smiling for the camera. Castelán now combines this "kinky touch" with explicit activism: the festival is said to be part of the "Alliance for Cultural Rights of Sexual Diversity and Non-Discrimination."

Yet the festival is more institutionalized than ever. The opening film is the transgender documentary feature *Quebranto* (which I treat in chapter 3). It was directed by Roberto Fiesco, who was by now the director of the Ariel awards ceremony, Mexico's equivalent of the Oscars. And Mix's short films are now coproduced by the festival with IMCINE, the national film institute. Sponsors now include the Spanish government (there is a sidebar of titles from that country) and the Marriott Hotel chain, in addition to gay venues around the capital. And striking out from the comfortable residential south of the city where the Cineteca is located, the festival now has a site in Cinépolis Diana, the most central multiplex on the grand avenue of Reforma (handy also for the gay Zona Rosa), a theater that is run by Mexico's most commercially successful exhibitor.

This process of authorization or consecration is revealed also in press coverage of the festival's journey from minor to major cultural status over the years. The first point to make is that, in spite of the event's early arrival, which predates the fall of the long-lasting and censorious PRI (Institutional Revolutionary Party) regime, there is no negative or homophobic press on Mix in the extensive print archives held in the Cineteca. Three critics are especially valuable as examples of diverse, but complementary, journalistic discourses on the festival.

First comes Carlos Bonfil, a respected openly gay critic for Leftist daily *La Jornada*, who founded a queer cultural supplement for the paper. In 2004 Bonfil praises the festival for its pulling power (the Cineteca's theaters, he says, are rarely so full) and its continuing success in promoting foreign queer auteurs such as Todd Haynes and Bruce La Bruce who were previously unknown in Mexico. Mix has become, Bonfil writes, a "reference point" for the capital's cinematic culture and a "symbol of resistance" to any remaining institutional temptation toward homophobia or censorship (Bonfil 2004).

For 2006 ("Mix Feliz"), Bonfil cites Castelán on "happiness" as the current queer condition and notes himself the happy coincidence that the festival began at the same time that medical progress meant that HIV was no longer a death sentence. For ten years, he writes, Mix has chronicled the "advances" for gays in Mexico, overcoming institutional homophobia, religious intolerance, and social discrimination. And it has offered a "parallel" history of the works of Mexican directors interested in "sexual diversity" who lack the option of commercial distribution. Among the highlights of the decade, Bonfil cites Mix's support for budding auteur Julián Hernández whose second feature *El cielo dividido* ("Broken Sky") opens this year's festival (I treat it in chapter 2).

Yet, as an ethically serious critic, the rigorous Bonfil still has criticisms. He complains that the festival has shifted over the decade from taking a radical position on sexual politics to embracing a diversity of options that include commercial cinema, explicit sex, and frivolity. Lesbian-themed features have also lost ground to "a festive parade of male nudes, directed toward gay self-consumption." This shift is dependent, however, on changes in the international circuits of gay filmmaking, where social conditions for queers are much improved, thus making activism appear less necessary than it was in many countries. And Mix's recent note of "optimism" is based also on its successful survival in Mexico against all the odds (Bonfil 2006).

It is an optimism confirmed in 2014 when Bonfil proclaims Mix's "coming of age" with the eighteenth edition. Bonfil cites a recent speech by Roberto Fiesco at the Ariel awards ceremony in which he said that not so long ago senior figures in the Mexican film establishment openly scorned "maricones" ("queers"). Now, writes Bonfil, Mix is "conquering ever wider audiences." Yet if the festival has helped to combat a homophobia that is no longer acceptable, its programming is, strictly speaking, "uneven in terms of artistic quality," owing to the urgency of its continuing quest for novelty. This is a sign, still, of immaturity. For Bonfil, then, the social good of queer integration may have been bought at the cost of lost political and artistic rebellion. Still, he is consistent in his support for the festival and its tireless director (Bonfil 2014).

Jorge Ayala Blanco is a yet more consecrated critic than Bonfil. The main reviewer in the business daily *El Economista*, he is also the author of numerous books and the recipient of a prize from the Mexican film academy. The award was highly controversial, however, given that Ayala Blanco is often vociferously hostile to much Mexican cinema and thus anathema to the local industry. It is highly significant, then, that like Bonfil, the normally censorious Ayala Blanco has consistently supported the Mix project.

Appealing to a typically idiosyncratic style (the titles of his books follow the letters of the alphabet), Ayala Blanco builds his reviews around motifs difficult to understand or translate. Thus (in 2006) he treats three features that are "resguardando" (protecting or sheltering) their diverse subject matter, and in 2004 he appeals to "pálpitos" (feelings or hunches) about his chosen films (2004a). In 2004 he also claims to find the "rhetorical-poetical figure zeugma" in three films once more. In 2008 it is the turn of "orientando" (directing or guiding) (Ayala Blanco 2008). While all of these terms are tenuously related to the films themselves, they serve, like Bonfil's more modest and accessible reviews, to lend discursive authority in the quality press to a festival that might be considered marginal. By taking even the most sexually graphic titles (and Ayala Blanco seems particularly attracted to these) and converting them into aesthetic objects in his reviews, the critic consecrates queer film, translating its audiovisual text into a rarified print idiom in keeping with the title of his column: "Cinelunes exquisito" ("Exquisite FilmMondays").

Finally, we come to Wenceslao Bruciaga, one-time gay correspondent of listings magazine *Time Out México* and later music and nightlife correspondent for Vice media website. Bruciaga's nonspecialist preview of the seventeenth festival in May 2013 (which includes a brief interview with director-entrepreneur Castelán) bears the title "Luces, cámara, ¡acción gay!" ("Lights, Camera, Gay Action!"). It is illustrated with a publicity shot of a pile of eight intertwined bodies, Castelán's young "stars" of film and dance (four will play parts in films and TV shows that I analyze in this book).

Unsurprisingly, Bruciaga's playful history of Mix focuses on its inextricability from sexual activity. Writing (unlike my previous critics) in the first-person plural, he notes that when the festival started in the 1990s there were fewer T's appended to the LGBT acronym (that question of labels once more) and the Internet was not yet a "neuronal appendage" to gays. "We" were thus "hysterical" if a date stood us up in the VIPS restaurant by the Angel of Independence monument on Reforma (not far from Cinépolis Diana). And "we" used to go cruising in porn cinemas in the Historic Center that have now been torn down.

When Mix began, writes Bruciaga, the capital's straight society (Bruciaga uses the gay slang term "buga") was only just starting to be accepting of queers. And gay events were no longer limited to the annual Pride march. It is no exaggeration, he writes, to say that Mix has contributed to the creation of "an authentic coexistence in diversity" (the phrase is familiar from the texts of the festival itself). And, no longer just a platform for exhibition and distribution, Mix has now ventured into production and the creation of audiences. Moreover, a new sidebar called "Transméxico" will be devoted for the first time to trans identities in the country.

Bonfil gives a scholarly account of the institutionalization of the transmediatic Mix within Mexico's cinematic establishment. Ayala Blanco, on the other hand, transforms crude homoerotics into exquisitely penned aesthetics. Bruciaga, younger and more reckless, travels the opposite path, conjuring up an urban sex tour for his complicit readers. It is one already hinted at, as we have seen, by changes in the festival's publicity materials such as the overtly sexy video *Vestido* with which I began this chapter. Bruciaga reinscribes the cinematic in the sexual and returns the officially authorized event to the marginal libidinal practices (such as public cruising in movie theaters) from which it had sought to extricate itself in its home base, the cinephile temple of the Cineteca.

Yet it might be argued, following the lead of Bruciaga's unapologetic first-person account of an erotic odyssey, that it is precisely by having the confidence to return to its erotic origins that Mix proves that it has successfully defeated the homophobia that scorned "maricones." By this sexual turn it demonstrates also that the festival has not abandoned a radical libidinal investment that even the international queer cinema circuit might prefer to disavow. Perhaps, then, and in spite of respectable appearances, the older members of Mix's devoted audience still remember how once they went to movie theaters not to take in esteemed foreign films but to cruise anonymous local bodies.

Web Romance: *Al final del arcoíris* ("End of the Rainbow," dir. Armando Silva Baena, Tres.Tercios/Contempovideo, 2008/2010)

¿Cuál es tu color para amar? Primera temporada: Azul

Ciudad de México. Armando acaba de salir del clóset y está cursando la preparatoria, la vida le tiene reservadas muchas sorpresas, él solo desea amar. Sus grandes amigos son su inspiración para seguir viviendo, Ana, la lesbiana, y Beto el afeminado del grupo y odioso de la prepa. Todos ellos vivirán una odisea por ser gay en una ciudad mayoritariamente hetero. En ese transcurso, Adrián, el gay confundido y Karla, la princesa del cuento, se les unirán para reírse, amar y descubrir lo que el azul les tiene deparado. La primera temporada AZUL tiene 16 capítulos que harán que te rías, emociones, ames y sientas. ¿Qué existe al final del arcoíris? El sueño ha comenzado. . . . ¿Cuál es tu color para amar? (original punctuation preserved)

At the End of the Rainbow. What's Your Color for Loving? First Season: Blue

Mexico City. Armando has just come out of the closet, he's a high school student, life has many surprises lined up for him, he only wants to love someone. His great friends are his inspiration to go on living, Ana, the lesbian, and Beto, the effeminate one in the group who hates high school. All of them will travel on an epic journey because they're gay in a city that's mainly straight. Meanwhile Adrián, the confused gay and Karla, the fairy tale princess, will join them and find out what the color blue has in store for them. The first season, called BLUE, has 16 episodes that will make you laugh, be moved, love, and feel. What is there at the end of the rainbow? The dream has begun. . . . What's your color for loving?

While, as we saw, the Mix Festival became institutionalized, in step with the increasing acceptance of homosexuality in cultural circles in the capital, my next text remains little known and truly marginal. *Al final del arcoíris* is a teen webseries produced by indie Tres Tercios, first distributed online in 2008 and released on DVD in 2010. Unlike even the most minor Mexican movie or TV show, it has no entry on IMDb and its fledgling actors were, to the best of my knowledge, never seen again (although they became minor celebrities in Mexico City's gay bars). The show proclaims itself "the first gay series in Mexico," a valiant and defiant aim. Yet it qualifies this claim to artistic novelty and creativity by stating inside the DVD cover that it is both "independent and amateur."

Like the festival, the webseries is a hybrid phenomenon. On the one hand, it cites mainstream telenovela both explicitly (with references in the dialogue to Televisa's straight teen juggernaut *Rebelde* ["Rebel," 2004–6]) and implicitly (with gay crushes replacing the more common passions of hetero heroes). But, on the other hand, it invokes documentary in its use of authentic, urban locations (especially exteriors), shaky cinematography, and grainy sound and video quality. And if the festival swung, as we saw, between the cultural and the sexual, *Al final del arcoíris* oscillates between the utopian and the everyday.

Thus, the ten central characters (only five are named in the synopsis above) live in a world wholly free from the restrictions imposed by the parents and teachers who never appear on screen. This is a major plot departure from conventional teen telenovelas such as *Rebelde*, which, in their quest for the family demographic, devote equal time to plodding plotlines for the adults that are surely scorned by young viewers. The high school kids of *Al final* are free to indulge in their search for same-sex love (three of the principals are lesbian) without distraction from their elders. Nor is there much evidence of homophobia here. Even Ana who has been thrown out of the house by her hostile provincial parents is given money by those same parents to set up home on her own in the capital (soon she will be joined by her less sympathetic gay brother). And, in an unlikely plot device, the young cast members frequently bump into each other on the streets where they live, somewhat implausibly in a megalopolis of over twenty million (mainly heterosexual) people.

Yet the unnamed *colonia* where several of the kids seem to be based, with its modest homes and yards, is blatantly everyday. *Al final*'s setting thus rejects both the flashy glamor of telenovela and the gritty horror of social realist festival films. Moreover, the scourges of the latter (drugs and alcohol) barely figure here. And the physical types of the amateur actors are likewise in-between: relatively attractive to fans of their own age (the characters are seventeen, the actors said to be "over eighteen"), none has the flawless look of mainstream Mexican media stars. The main character Armando (named for the indie entrepreneur who wrote and shot the series) has somewhat uneven teeth; Ana is a little overweight; her brother is (already) losing his hair; the "confused" Adrián is skinny and ratty. The target audience can thus aspire to the everyday status of these queer kids. Viewers can also follow the young people more closely on the web than they could on the big or small screen, as the ensemble cast initiate a halting journey toward lesbian and gay maturity beset by romantic obstacles that are all too commonplace.

Romantic is the word here. While Mix promotes self-consciously kinky eroticism amid the high culture, *Al final* suggests that love, not sex, is all you need. Indeed, Armando will lose his first boyfriend, victim to a hit and run accident on a city street, before they have done more than share a brief kiss. By contrast, when campy Beto is fucked by a hunky plumber in his own home, the scene reads like a porn-fueled fantasy. And it is something of a shock when we discover that it is meant to be real. Beto confesses to his friends that he is no longer a virgin (they chide him for not using a condom).

Yet in spite of its general lack of explicitness (we are offered just a few shots of rear nudity in bedroom or shower scenes), *Al final* tackles themes little explored in festival films. While Mix included fewer lesbian titles over the course of its run, the webseries gives prominence to Ana's slow wooing of an initially straight friend, Karla, in spite of the hostility of Ana's still enamored ex. A parallel sentimental journey takes place between the "confused" Adrián and the decidedly nonconfused, effeminate Beto, who always speaks of himself and his friends in the feminine gender. And it is especially surprising and gratifying to see the sober, nominally straight boy pair off with the screaming queen. *Al final* thus takes sissies seriously. Effeminophobic statements are voiced only by Rodrigo, a straight-acting closet case, who is hardly a sympathetic or trustworthy spokesman.

One important theme of the festival is indeed taken up by the series, namely, labeling. And a vital aspect of *Al final*'s everydayness is its vernacular language. It is no surprise that the characters use the same kind of gay slang we saw in the *Time Out* article by journalist Bruciaga, such as "buga" for "straight." But they also comment on mainstream terms. Protagonists Armando and Ana define themselves explicitly and unapologetically as "gay" and "lesbiana." But when "confused" Adrián uses the word "homosexual," Beto laughs, saying he sounds like uptight Charlotte in *Sex and the City* (HBO, 1998–2004). And Beto brands Adrián, his future lover, "hetero-curious."

The kids often use Anglicisms in their speech ("Too much information!" exclaims lesbian Ana when Beto narrates his first sexual experience). But in spite of these media and linguistic references to the United States, the kids' language and behavior are still firmly based in Mexico. Ana sees herself as a provincial (her family remains in distant Michoacán) even though she lives successfully alone in the megalopolis. In an early episode, grungily shot and muddily recorded, all the kids make a first visit to the Zona Rosa gay village. Standing on iconic Calle Amberes, an easily recognizable pedestrianized street, they give each other lessons in getting the phone numbers of other teen passersby. And, under the Angel monument on Reforma, Armando almost scores a kiss from the boy he has been flirting with. Beto and Adrián later get together while sitting on a distinctive bench in the shape of a pair of hands on that same grand boulevard. And when they walk off together hand in hand, among doubtless bemused tourists and locals, the show signals the entrance of youth homosexuality into the everyday life of the capital. It does so in an engaging and optimistic way that is far indeed from the miserabilism of many of the festival films shown at Mix.

The last episode of the first season ("The Best of All Possible Endings") is somewhat more ambivalent. Daringly, it consists of a single sequence set in a single location: lesbian Ana's birthday party is being held in the small garden behind her house, which is decked for the occasion with pink balloons. Stressing the tension between community and individuality, the documentary-style hand-held camera work alternates between crowded group shots of the assembled teens and tight close-ups of our now familiar fledgling stars. Exploiting depth of field, one early setup has a festive group in the foreground disrupted by the appearance of a looming presence in the background: black-clad bad boy Rodrigo, who clutches a blister pack of illicit pills, which he proceeds to swallow.

Cutting between close-ups on handsome, spiteful Rodrigo and two-shots of him with other more festively dressed cast members, the extended dialogue has him insult each friend in turn for supposedly having abandoned him. When he angrily stalks off, the camera follows the formerly "confused" straight-identified Adrián and extravagantly campy Beto (here in vivid red tank top and golden party crown) out into the street (they had recently broken up). As Beto sits on his moped, Adrián declares his love once more, telling him not to be afraid for the future. They smile, hug, and publicly kiss (several times). It is a long take in two-shot, signaling the renewed commitment of a touching but unlikely couple.

Charmingly, at the end of this sequence shot, as the new couple walks off hand in hand, the actor playing Beto realizes he has left the keys in the moped and goes back to retrieve them. Such vérité moments of apparent immediacy are reinforced by the final sequence of the episode, a black-and-white montage of the cast's past encounters, both pleasurable and painful. The voiceover states that over the thirteen months of the season the characters have changed from "trembling little chicks" to adults who know "what it means to be gay and to form part of a community." Addressing the viewer directly, it tells him/her: "You are not alone." It is a close connection with the target audience that is facilitated by the artfully casual shooting and endearingly amateurish performance style of this unique webnovela.

In a rare article on Mexican American webseries, Tomás López Pumarejo (2013) sketches a few characteristics of this still marginal genre (Juan Piñón also treats the genre, its branding, and interactivity in 2014). López Pumarejo suggests official series made around the time of *Al final* by Univision (Televisa's US collaborator) are overtly based on a product placement that is more explicit than in broadcast telenovelas and are tightly tied in to the star system of mainstream TV (López Pumarejo 2013, 310). The runs of webnovelas are much shorter (just fifteen episodes to the telenovelas' 150) as is the duration of each episode (five minutes compared to forty-five) (311). And there is another major difference between the two: the webnovela seeks to engage active users, not to entertain passive viewers, even as it remains strictly conventional in format.

Typical of López Pumarejo's traditionalist series (and still available on YouTube) is *Vidas cruzadas* ("Intersecting Lives," 2009), which stars Kate del Castillo, later notorious for her controversial visit with Sean Penn to criminal kingpin El Chapo. Here Kate's auburn mane is shown to be due to L'Oréal hair color, which gets its own shameless close-up in the very first episode. Product placement is less emphatic in the low budget *Al final*, although one lengthy sequence has Beto getting a new hairstyle (spiky highlights) in a real-life salon whose name is prominently featured in that episode.

But I would read the use of such trademarks here as part of *Al final*'s youthful pedagogy of the everyday. While few Mexican housewives could aspire to the life of Kate (*Vidas* is set in Los Angeles), queer teens in, say, provincial Michoacán might well imagine themselves warmly welcomed into a gay-friendly hair salon in metropolitan La Roma. And they could, no doubt, cruise like the complete cast in the Zona Rosa, kiss under the Angel like Armando, and stroll hand in hand along Reforma, following the brave, tender example of Beto and Adrián.

A stills gallery on *Al final*'s DVD has the large cast posing and kissing in their branded leisure wear on a city roof. Flagrantly commercial, the shoot is also (like the young stars' performance style) endearingly amateurish, producing a potent reality effect that surely seduced young viewers or, in López Pumarejo's word, "users." Indeed, each episode (which runs for not five but a relatively extended fifteen minutes) ends with a blooper reel of the teen stars fluffing their lines and clowning for the camera, offering faithful viewers the promise of direct access to new queer youths, new queer lives. Within the context of the nascent transmedia genre of *webnovela*, *Al final* is highly innovative, and not just for its unique gay content.

As we have seen, then, *Al final* is also hybrid, drawing both on mainstream melodrama (that tragically early death) and indie documentary (those murky scenes with a muddy soundtrack taped on city streets). Where the Mix Festival queered Diego and Frida, *Al final* homosexualizes telenovela and cinéma vérité. And, like Mix once more, *Al final* is embedded in multimedia. Thus, the webseries is based on a print novel, less known even than the series. And it is promoted in Tres Tercios's gossip magazine called *Homos*, a gay version of Televisa's *TVyNovelas* that is accessible on the web. The same gay company has also edited a more explicit erotic publication featuring local models called *DFÑOS* ("MexicoCitizens") and posted a regular podcast.

All of these projects are created by a minor entrepreneur as indefatigable as Mix's Castelán: Armando Silva Baena. Pictured on the cover of *Homos*'s issue for October–November 2014, he shares billing with Elton John and Lady Gaga. Inside he is given a lengthy interview and lavish picture spread in which he poses casual in jeans or formal in tuxedo, at one point sporting an ironic royal crown. The article recounts Silva Baena's odyssey over six years as a unique gay media producer.

The interviewer acclaims him as "the person who made it all possible," "the creator of an internet dream factory," and "the pioneer of the gay entertainment industry in Mexico" (Legnar 2014, 20). And his self-related career journey begins with reality shows ("a search for new talent") before he tried his hand at drama series (30). International distribution on the web and national sales at Mixup DVD stores (where I came across my copy) were vital. But the secret of *Al final*'s success was, he says, rather the relative lack of gay content back in 2008 and the "sincerity" of novice producers who worked "with their hearts." This honesty, he claims, was readily recognized by fans. Silva Baena says also that the "media phenomenon" of *Al final* did not change his own life but did transform those of his young cast who were "recognized on the street and chased after in [gay] clubs."

But not everything was rose colored in Silva Baena's rainbow-hued homo world. That same cast later took him to court. And a second season boasting some twenty actors was less successful. However, *Al final* remains for him "my story, my love of the rainbow, my discovery of the color for loving, my first challenge that was fulfilled and achieved" (Legnar 2014, 31). Entrepreneurship has rarely seemed more solitary than here. And the interview does not skip over the personal disappointments that punctuate a professional career. The optimistic *Al final* was, we are told, born out of a "deep depression" (29).

Yet all of Tres Tercios's multimedia projects are addressed directly to the user or consumer. Hence the series' tagline "What is your color for loving?" Silva Baena ends the interview by effusively and repeatedly thanking "you," his unseen audience. And just as the Mix Festival, recently acclaimed as a milestone in Mexican cultural life, sought to conjugate the artistic with the erotic, so Tres Tercios attempted, in its brave, unrecognized way, to combine the utopian search for love with the social concern for everyday gay life in the city. It is a potent cocktail that will also be found, surprisingly enough, in the porn producer Mecos.

Porn Stories: *Corrupción mexicana* ("Mexican Corruption," dir. El Diablo, Mecos, 2010)

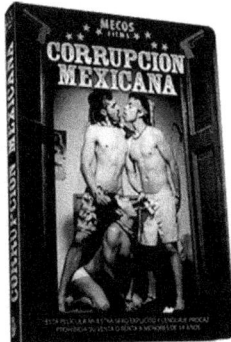

Una audaz producción que nos lleva a un viaje por el país a través de sus calientes y corruptos habitantes. Cuatro historias revelan la cachonda realidad de México: secuestros, soborno, violencia, sexo rudo, humillación, drogas y engaños. . . . Cualquier coincidencia con la vida real, no es casualidad . . . es pura calentura!!!

A daring production that will take us on a journey through the nation by way of its hot and corrupt inhabitants. Four stories reveal the horny reality of Mexico: kidnappings, bribes, violence, rough sex, humiliation, drugs and deceptions. . . . Any similarity to real life is no accident . . . it's just hotness!!!

If homosexuality is (or was until recently) in itself marginal and unauthorized in the context of Mexican film festivals or webseries, gay pornography would appear to be doubly scapegoated, combining as it does a minority sexual preference with a widely despised film genre. Yet, of the three phenomena I treat in this chapter, only Mecos has been dignified by academic study both inside and outside Mexico.

So-called adult film has long been institutionalized as a specialism in US media studies, although less commonly in Mexico. Like film festivals, it benefits from an established Scholarly Interest Group at the Society of Cinema and Media Studies, which hosted no fewer than ten panels on the subject at its annual meeting in 2015. The academic quarterly *Porn Studies* was launched in 2013. And as long ago as 1989 Linda Williams had vindicated heterosexual porn as a rare genre in which women were not objectified but rather presented as the active subjects of the film narrative, however vestigial.

What I argue here is that (like my previous objects of study) Mecos's productions are hybrid (combining the erotic and the social, the utopian and the everyday, documentary and drama); transmediatic (fusing video, Internet, and piracy); and entrepreneurial. Mecos, still the only producer of its kind active in Mexico, was founded in 2004 by a minor auteur known as "El Diablo," who is as tenacious in his own field as Mix's Castelán or Tres Tercios's Silva Baena are in theirs.

Mecos has also benefited from support from Mexican film scholars and practitioners. As early as 2009 cultural journalist Sergio Raúl López (a contributor to educational TV Channel 22) and producer Roberto Fiesco (producer of Julián Hernández's films) published a double article on Mexico's first gay porn productions in *Cine Toma*, the official journal of the Mexican film academy and film school, itself a division of the UNAM, the capital's national public university. Fiesco praises *SeXXXcuestro* ("XXX Kidnap," 2002), which he calls a "media milestone," for opening up a debate on Mexican pornography. This is all the more surprising because of the film's gay theme in a country that is "traditionally homophobic" (López 2009, 32). Like Fiesco himself, the pseudonymous photographer and director (to whom Fiesco attributes later titles by Mecos, although the name given elsewhere is different) were graduates of CUEC (the "University Center of Cinema Studies"), one of the two official film schools. Yet, Fiesco writes, the fledgling porn producers experienced the same problems of the creators of any Mexican film: difficulties with exhibition, unfair demands from the distributors, and problems in reinvesting profits in future projects (32).

The two veterans of *SeXXXcuestro* give Fiesco a detailed account of the production process, including casting and shooting, through which they intended to make a "good production." While they sought a reality effect (the director says that he enjoys "non-acting" [López 2009, 33]) the shoot proved as complex and contrived as any other, with the performers (provided by an escort agency) needing careful coaching to achieve the required cinematic effect. Even the editing was fudged, with "money shots" recorded later edited into original footage that was filmed in over just two days in an abandoned restaurant (34).

Conscious of a national context in which their production was unique, still the producers had an international dimension in mind. While US videos lacked narrative, the Mexicans sought in their own work a new rhythm and format, complete with an edgy story and dialogue (real life kidnapping was not yet as sadly frequent in Mexico as it became in later years). They received no homophobic treatment from the Mexican state, which even allowed them officially to name their company for tax purposes La Verga Parada (The Erect Cock) (López 2009, 36). Meanwhile, their film was shown in the Toronto Gay Film Festival. And they say they were well aware of French art movies, such as those of Cathérine Breillat, which incorporate hard core footage. In the hands of such cinephile producers, then, porn is not so far removed from more legitimized media practices and genres.

Sergio Raúl López's interview in the same issue of *Cine Toma* is with El Diablo, the founder of Mecos. The company's first film is said to be *La putiza* ("The Beating," 2004), described as "a crazy gay porn flick that takes place among mariachis, Aztec masks, and lustful wrestlers" (López 2009, 35). Like its predecessor *SeXXXcuestro*, *La putiza* was shown abroad at specialist festivals in Barcelona, Toronto once more, and Philadelphia. It was followed by another wrestling movie, *La Verganza* (2005, the title is a pun on "cock" and "vengeance") and three reality-style audition compilations *Selección Mexicana 1*, *2*, and *3* (2008).

As a successful entrepreneur, El Diablo proves well aware of current conditions of production, distribution, and exhibition for his titles. Thus, he submits novice performers to a process of "professionalization" for works that are marketed in the three windows of Internet, DVD, and piracy (López 2009, 35). And once more the home/abroad dialectic is vital. Local consumers seek work they can enjoy for once "in their own language, with their own people"; in Japan and the United States customers desire rather something very different from the gay porn made in their own countries (36).

Yet El Diablo also muses more conceptually on his chosen genre, discussing the tenuous boundary between the erotic and the pornographic (López 2009, 35), one of the key themes of this book, and claiming finally that "for me sex is connecting with the earth, the animal . . . it's so huge that it represents a great responsibility" (36). El Diablo's investment is both commercial and libidinal, and even ethical and emotional.

It is a UK-based scholar, Gustavo Subero, who has written in detail about Mecos. Subero's main argument is that this unique producer deserves praise for its embrace of a full ethnic range of Mexican bodies and for its resistance to the implicit racism of US videos that feature only hyperbolically sexualized Latino performers. Subero focuses in his articles (2009, 2012) on the boxing movie *La Putiza* and the *Selección* audition films, respectively, and in his later book he treats *Corrupción mexicana* (2013,

179–87). Inspired by his pioneering work, I myself will treat the last two in my own rather different context of minor genres or auteurs.

The credits to *Selección mexicana* begin with a quick montage of postcard shots of Mexico City, including Reforma and the Angel monument, sites also prominent in *Al final del arcoíris* (two of Mecos's performers will even lounge provocatively on the same bench where Tres Tercios's non-actors filmed a more romantic encounter). Segments generally begin with interviews shot indoors or sometimes on the balcony outside Mecos's studio, which is situated on a quiet street in bohemian La Roma. The supposed novices are asked by a voice off screen about their origin (in the capital or provinces), their first time (which often seems to be with family members), and their reasons for making porn (often they have revealed their intentions to only a female best friend).

These micronarratives serve to establish what *SeXXXcuestro*'s director called a "non-acting" that surely heightens viewing pleasure and is analogous perhaps to the "sincerity" claimed by Silva Baena for the performers in his webseries. But they also individualize performers who would otherwise remain anonymous and objectified, permitting (like *Al final* once more) identification with as well as desire for subjects that embody "my language, my people" for local audiences.

It is striking that one handsome performer in the first feature-length film of the *Selección* trilogy, whose name is given as "Tomás," is smiling and animated here, willingly answering questions as his scene partner absentmindedly holds his penis. But when "Tomás" appears under a different name in Julián Hernández's hard core art short *Bramadero* (2007), he adopts throughout an expression of affectless alienation that is more appropriate to auteur cinema (we shall come back to this short in the next chapter). In spite of their common graphic sexual content, there is little doubt that it is Mecos's film that comes closer to its audience, inviting us to feel affection for the man as well as excitement and admiration for his exposed body.

Yet there is much variation between the segments in the *Selección* features. In the first scene (the one starring the *Bramadero* performer) the two conventionally handsome men with gym-trained bodies flip/flop fuck on a couch. In the second, more complex in structure, three younger, slimmer men engage in a more complex sexual repertoire, in which one of the two cameramen often comes into shot. Indeed, at one point the owner of the voiceover who is heard directing the action throughout (El Diablo himself?), visibly aroused also, participates in the scene. The third segment stars two extravagantly attired punks, with green and pink hair and multiple piercings; while in the fourth and last a muscular, darker skinned and mestizo-featured man penetrates a younger more European type. It is striking that some of these performers do not provide the money shots generally held to be essential to the genre, proof perhaps of a genuine improvisation on set.

Subero gives a lengthy and subtle account of *Selección*, whose title ironically mimics that of the Mexican soccer team. And he treats *Corrupción mexicana* ("Mexican Corruption"), which was released later in 2010, in his book. My own argument is that while the *Selección* films combine the utopia of sexual release (which director El Diablo describes as a "great responsibility" in itself) with the little narratives of the everyday, *Corrupción* embeds four sexual numbers within a loose narrative that invokes the most serious sociopolitical problems facing the nation. As the Mix Festival queers high art and Tres Tercios gays telenovela, Mecos homosexualizes kidnapping, drug trafficking, and anti-indigenous racism with a playfully defiant erotic eye.

In an opening sequence set in the city, two men (one the wrestler from *La Putiza*, the other a trainee from *Selección*) are forced to service the cop who has arrested them (he keeps his helmet on). In the second a "fresa" (entitled rich kid) is kidnapped by a criminal couple and obliged to undergo a double penetration (his father, listening in on the action on his son's phone, is also aroused, as is a friend later chatting with the abductee on Mecos's website). A cowboy-hatted villain then penetrates an Indian in a wheelbarrow, tricking him with a romantic promise of a life together in the capital (the camera lingers in a picturesque provincial town on a street called appositely The Sad Indian). (Subero does not mention this scene, called "Coger al indio" ["Fucking the Indian"], which might call into question his optimistic vision of Mecos's "multiculturalism and multi-ethnicity" [Subero 2013, 180].) And, finally, a soldier spots a youth urinating on the street and subjects him to a lengthier golden shower of his own.

It seems appropriate, given this apparently social realist setting, that, unlike the professionally buffed escorts of the similarly themed *SeXXXcuestro*, the performers in *Corrupción* are physically imperfect. The ex-wrestler is now losing his hair, the Indian is short of stature, and the dreadlocked kidnappers, who force a gun into the *fresa*'s mouth, are scrawny. Like the ephemeral stars of Tres Tercios's webseries, Mecos's good-enough-looking porn performers offer viewers a plausible opportunity for identification as well as for desire.

Erotic fantasy is here, moreover, inscribed in a highly charged and multiethnic context of crime, prejudice, violence, and exploitation. Mexican corruption, indeed. Yet rather than reading this move as a frivolous evasion of social problems, I see it as an exorcism or catharsis of genuine dangers that Mexican viewers, whether gay or not, have good reason to fear in everyday life. Just as Mix created a safe haven for eroticized film art and Tres Tercios produced a utopian world for a queer youth freed from parents and teachers, so Mecos, more radically perhaps, conjures away (temporarily at least) onerous social scourges that have hitherto eluded solution by Mexican authorities.

Indeed, the film suggests with its depiction of the police and the army that those authorities may well be complicit with the criminal acts recreated on screen. And the key motif here is once more the journey. There are many driving shots in *Corrupción mexicana*, sometimes sexualized (a passenger fellates a driver). Like the long trajectory of the producer Mecos, they suggest a certain movement of queer Mexican bodies in space and time, a movement that is at once pleasurable and profitable.

Loving the Real

In December 2015, I paid a visit to La Roma, where Mecos's studio, familiar from their films, is lodged in a typically ramshackle period home (the now middle-class *colonia* was hard hit by the earthquake of 1985). Although Mecos's website claimed that their in-house store would be open for sales (I was hoping to glimpse also their famous couch), no one answered when I rang the entrance bell. The Mix Festival now also boasts a venue in La Roma, albeit farther south, the trendy Cine Tonalá, which is known for its cuisine and social events. We remember that *Al final* too ventured out of the Zona Rosa to La Roma when campy Beto enjoyed an expert new haircut.

There could be no greater proof of the localization of the media practices that I have treated here and of their interconnection in the same urban space than these precise references to a neighborhood. And unlike the monumental Reforma or Angel, La Roma is recognizable only to cognoscenti. Beyond this insistent localism, however, Mecos's distribution remains the widest ranging. I bought my (legal) DVD of *Corrupción mexicana* in a Zona Rosa sex shop, its availability on the shelf confirming El Diablo's claim that Mexicans will still pay full price for copies in order to support the producer. But some of Mecos's titles could also be viewed at that time on Internet pirate sites, albeit in mutilated form, often shorn of the crucial interviews with the performers.

In January 2016 Mecos's back catalog suddenly became available on AEBN (Adult Entertainment Broadcast Network), an official US pay site. Strikingly, the Mexican company had now also collaborated for a self-titled DVD with the San Francisco–based label Treasure Island. On AEBN, Mecos films, which play at full feature length, have no subtitles. They thus force US viewers to translate "our language, our people" into their own idiom, which is no doubt part of the pleasure that curious North American viewers take in these rare Mexican-made titles. Silva Baena of Tres Tercios also spoke in interview of the importance of distribution of his series by both the physical stores Mixup in Mexico (where I came across *Al final* for the first time) and on international websites.

My three subjects remain fragile, however. An anodyne trailer for the second season of *Al final* (here the key color is not "Blue" but "Green"), which shows the cast clowning in a park, had provoked by January 31, 2016, mixed reactions in viewers on YouTube (*Al final del arcoíris* 2009). One sent congratulations; another asked why gays always had to be effeminate in Mexican series, claiming we are "not like that"; a third wrote that the series was denigrating to the "homosexual community" because of its poor technical quality; a fourth erroneously cited "hot sex" that was nowhere to be seen; while a fifth urged "Jehovah [to] take pity on these queers [*jotos*]." Such residual self-hatred and homophobia prove that surely there is a need still for gay media utopias in Mexico, whether they are artistic, romantic, or erotic in nature.

Social media accounts on Twitter and Facebook, where global gay culture is invariably hybridized with national references, reveal similarities between all three of my case studies in this chapter. And we have seen that even in Mexico, where the Internet remains less financially and technically accessible to a broad public than elsewhere, transmedia can connect or perhaps create active audiences over a continuous period of time. Yet queer content still renders unstable cultural legitimation in all genres and media. Distinction is thus perilously acquired by gay and lesbian feature film through the festival over the course of its run and is barely enjoyed by webseries or porn, however innovative these last two may be in their own national context and however respectfully they may be treated in occasional scholarly articles in Mexico and abroad.

On the other hand, my research shows that actors and directors may move with relative freedom (at least when they are disguised by porn aliases) between adult videos and escort sites and state film schools and art houses. In the next chapter, we shall see how even Julián Hernández, the most austere of gay auteurs, draws on my minor genres, practices, or discursive contexts. For now, it is enough to suggest that all these have surely shaped the audiovisual consciousness of queers in Mexico. This is the case even as they remain invisible to those foreign viewers and academics whose experience of Mexican LGBT content remains mostly confined to an exclusive art house circuit that bears little relation to local queer production.

THE ART CINEMA OF JULIÁN HERNÁNDEZ

Two handsome young mestizo-featured men embrace in silhouette in front of a window as a hazy view of the city can be made out far below. The camera pans slowly left over the bright walls, which are interspersed with patches of thick shadow. And, suddenly via a masked cut, we are in another room, facing a door. It opens to reveal one of the youths, accompanied now by a different, equally handsome but more Mediterranean-looking guy. They enter, smiling, and kiss on the bed.

Panning right, we see the first boy, now shirtless, examining himself in the mirror. A further pan, this time left, carries us back with him to the bedroom. The camera tilts down a little (still, apparently, all in one shot) to show the other youth, now lying frontally naked and seductive on the bed. He reaches out smiling to the first, beckoning to him with outstretched hand. On the soundtrack we hear, first, a dispassionate voiceover recounting a story of irrevocably lost first love, and, second, a romantic ballad, also of impossible and tantalizing passion, by José José, Mexican singing star of the 1970s. It is called "Tan cerca, tan lejos" ("So Near, So Far").

This is the final sequence of Julián Hernández's second fiction feature *El cielo dividido* ("Broken Sky") from 2006. With its attractive and amorous young male cast, bravura camera work, and lush soundtrack, it is highly characteristic of its director's subject matter, mise-en-scène, and cinematography. Indeed, a very similar scene (complete with the switch of one boy for another within a single panning shot) can be found in Hernández's third film *Rabioso sol, rabioso cielo* ("Raging Sun, Raging Sky," 2009). In the light of these similarities, one Mexican critic told me after the Morelia festival screening of Hernández's latest feature *Yo soy la felicidad de este mundo* ("I Am Happiness on Earth," 2014) that Hernández's work was now too familiar, too much the same.

Yet is a criticism that directors who confine themselves exclusively to heterosexual subject matter are unlikely to face. And, with four features and a large number of shorts, it is Hernández's unique achievement to have become (with the invaluable support of his producer Roberto Fiesco) the only openly gay career director in Mexico and the only filmmaker tout court in the country to consistently treat queer themes. Hernández's sole precedent would be the veteran Jaime Humberto Hermosillo, who currently occupies a respected, but marginal, position in the filmic field.

Such thematic consistency and formal continuity are of course characteristics of a director who achieves the distinctive status of auteur, as are the enigmatic film titles and extended running lengths also favored by Hernández. And his journey to the privileged but perilous status of queer art house director is inseparable from the Mexican festivals that first show his films. Thus, Arturo Castelán, director of Mix, was an associate producer of *El cielo*; and Alan Ramírez, the boy in the frock from Mix's *Vestido* video, starred in *Yo soy*.

Yet, as we saw in the previous chapter, the distinction conferred by LGBT festivals is fragile. And writing on "The New Homosexual Film Festivals" in 2006, the year of *El cielo*'s release, pioneer LGBT theorist B. Ruby Rich both cites Mexico's Mix as the "powerful seedbed for local production" of "exciting queer films" (including those of Hernández) (620) and quotes criticism coming even from queer constituencies that such festivals are no longer needed:

> Critics contend that choices have become available enough in the mainstream to obviate the need for such specialization; that queer audiences have been transformed into a niche market no longer in need of "ghettoized" events; that the [LGBT] film/video festivals are a stale holdover from the early post-Stonewall era. (620–21)

Given the belatedness and unevenness of Mexico's experience of sexual liberation (not to mention the lack of a Mexican equivalent of Stonewall), the argument would not seem to hold true for Mix, which remains much needed. But, if we turn to the US trade press, we see how Hernández's artistic career and friendly foreign reception have been inseparable also from participation in his country's most prestigious generalist festivals.

Thus, when *El cielo* premiered in competition at the Guadalajara International Film Festival (it was to win nothing), *Variety*'s Anna Marie de la Fuente (2006b) placed the twenty-first edition of Mexico's largest, oldest, and most prestigious film festival in context. After years of decline in quality and organization, the event was looking to improve its reputation with the appointment of experienced and respected producer Jorge Sánchez as its new director. Yet the Mexican national focus of the selection had already been diluted when in 2004 the festival's competition had been broadened beyond local films to include for the first time Latin American and Spanish titles. And in 2006 Spain was the first nation invited by Sánchez, who had a track record as the former head of the international group the Iberoamerican Federation of Cinematic and Audiovisual Producers.

Moreover, if, as Sánchez is quoted as saying here, cinema has a "double nature . . . as art and as industry" (De la Fuente 2006b), then the film facts reproduced in the article are contradictory for the Mexican film business. Thus, on the one hand, box office and screens had increased in 2005 when compared to the previous year, and Mexican films released had also risen (from eighteen to twenty-three). But, on the other hand, since production had also gone up to fifty-three, this meant that thirty new Mexican films had not reached the screen. Meanwhile box office share for Mexican cinema as a whole had fallen from 5 percent to 3.5 percent, and the average box office take per Mexican film had collapsed by 44 percent. This was not, then, the most propitious time for a marginal director like Hernández, working on microbudgets, to premiere *El cielo*.

Conversely, John Hecht (2006) wrote in the *Hollywood Reporter* the same year that "buyers [saw] new relevance for Guadalajara film mart," claiming it had become "a key stop on the film festival circuit." Hecht's first example here is also *El cielo*, whose world rights were acquired by distributor Fortissimo Films and which is described as "a gay-themed drama co-produced by Mil Nubes Cine [Hernández and Fiesco's own company, named for their first feature], IMCINE (the National Film Institute), and Mexico City film school CUEC." Three years later, when *Variety*'s Robert Koehler (2009) hailed a "bumper crop of Mexican features" at the festival, he also praised positive industrial changes: the "bustling film market," which is linked to the one in Cannes, the Talent Campus (linked to Berlin), and "Mexico's landmark corporate tax incentive for film production" which had passed three years earlier. In this edition of the festival Hernández was premiering, in competition once more, his longest and most ambitious feature, *Rabioso sol, rabioso cielo*.

Yet, although Hernández surely benefited from these international connections (he had won Berlin's Teddy award for his first feature, followed by three Ariels, or Mexican Oscars, in minor categories), *Variety* once again saw "gay-themed pics" as a "tricky sell . . . in macho Latin America" (De la Fuente 2006a). Ang Lee's cowboy romance *Brokeback Mountain* (2006) was being promoted in the continent on the basis of its "quality" and "awards," not its subject matter. Once more Hernández is the sole example given here of a local queer director. *Mil nubes* did better business abroad than at home, and distributors hope that *Brokeback* "will help boost acceptance" of *El cielo* in what is described as a traditionalist "religious Mexico."

By 2014, however, when *Yo soy* opened in the Morelia International Film Festival (the smaller, artier rival to Guadalajara) Hernández is described by James Young in *Variety* as a "heavy hitter." Once more he benefited from an institutional change: previously the festival had accepted only first and second films from Mexican directors, and this was his fourth. Hernández also spoke at a moving and informative round table, discussing the now-distant origins of the Morelia festival as a showcase for Mexican shorts by novice directors such as he. The gala screening that year at Morelia was

Alejandro González Iñárritu's *Birdman*, later to win Oscars for best film and director, which was viewed by *Variety* as "lead[ing a] strong selection of homegrown fare."

Hernández, however, sought to distance himself from the trio of internationally feted Mexican amigos (Iñárritu, Alfonso Cuarón, and Guillermo del Toro) who of course enjoy much bigger budgets and wider audiences than he. He was quoted by the States News Service ("Mexico Fetes Cuarón's Oscars" 2014) as saying: "These three do not make Mexican film. They do not make their films in the Mexican system and their themes do not result from living here in the society where the rest of us live." The anonymous reporter notes that Hernández's "brooding, homoerotic films have won international awards and foreign distribution, but have seen little commercial success in conservative Mexico."

In the trade press and on the festival circuit at least, then, Hernández had over the course of a decade managed the considerable feat of plausibly representing a country whose general audiences did not seem to have warmed to his films. Moreover, a filmmaker whose aesthetic seems at first sight noticeably abstract (those balletic pans and long takes) and whose subject matter was specialized ("gay-themed" and almost exclusively Mexico City–based) staked a recognizable claim also to depicting everyday life in the home country where he, unlike some of his more favored colleagues, had continued to make his career.

Unsurprisingly, academic commentators on Hernández tell a different story about art and industry. And two major scholars (Gustavo Subero and Laura Podalsky) have differing readings of *El cielo*, both subjective and objective, which I treat in the next section of this chapter. To conclude this introduction, however, I will present my own intertexts to Hernández's unique body of work. They are a sociological account of gay Mexican lives based on informant interviews and a revisionist reading in a new queer light of the fetish figures of Mexican cinema of the Golden Age. Between the two of them, these works offer invaluable micronarratives of present and past (in life and in film) that we can go on to juxtapose with Hernández's gay cinematic subjects.

Héctor Carrillo's *The Night Is Young: Sexuality in Mexico in the Time of AIDS* (2002) deals with a different city and decade from Hernández's features, focused as it is on provincial Guadalajara (the home of the festival mentioned above) and on the 1990s (when Hernández had made only shorts). Yet, with its detailed case histories of love and passion, Carrillo's study, at once anthropological and sociological, offers an interpretative context for Hernández's fiction films, which focus on the same theme.

Carrillo's title is based, characteristically, on an anecdote. When he, a US-resident Mexican, proposed in a Guadalajara club going home at midnight to rest, the gay friends with whom he was spending the evening were amazed by his suggestion. For them, writes Carrillo:

> The night was constructed as the time of play and socialization, and thus the time of things sexual . . . there was an explicit acceptance of transgression of daytime rules and of exploration of the veiled prohibitions exerted in "normal," everyday life. The night was the time when sexuality made its full appearance, taking center stage and permeating all the interstices of social interaction. (x)

Carrillo follows up this initial sketch of Mexican cultural (sexual and social) expectations with a first case study, that of "Antonio" (all names are pseudonyms), a young businessman who works in a fancy office tower (Carrillo 2002, 1). Antonio identifies as "gay," using the originally English word that is now variably adopted into Spanish. Like some of the performers in Mecos's audition videos we examined in chapter 1, he was introduced to sex by being penetrated by members of his own family. But he also likes to seduce and penetrate male co-workers whom he identifies as "heterosexual" (2). And he is himself "infatuated" with a male lover who penetrates him (3). In this latter relationship, Antonio does not ask his lover to use condoms, as he feels "the need for love" is incompatible with the dispassionate negotiation required for safe sex.

Antonio considers himself politically "moderate," saying "I like Mexican tradition." Yet he is, Carrillo writes, "well aware of a contrast between traditional Mexican ideas about sex and 'newer' ideas that were being adopted, including his own ideas about homosexuality and gay identities" (Carrillo 2002, 3). Such internalization of social change does not, however, prevent Antonio from effortlessly playing the part of the macho when dealing with his female secretary. It is perhaps no accident that, as signaled by his earlier reference to sex "taking center stage," Carrillo's key theoretical concepts are dramatic or ludic. They are "cultural scripts" ("the syntax and understandings of roles and performance") (5) and "strategies" ("individuals' understanding of the 'rules of the game' . . . and how best to participate in the 'game' of social relations") (6).

Carrillo's many detailed case studies, with their individualized scripts and strategies, problematize the well-known sex/gender model of sexual identity categorization still suggested by many scholars of Mexico. Carrillo presents in a figure the three positions according to this schema, giving their Mexican names: *hombres/hombres normales* (masculine acting and identified, attracted to women and/or men and assumed to be inserters in sex with men); *internacionales* (a term already falling out of use, referring to men who are both *activos* and *pasivos* in sex with other men); and *maricones* (effeminate males, assumed to be exclusively attracted to men and acting as *pasivos* with them)

(Carrillo 2002, 38). The newer, international object choice model, now also operative in Mexico and shown in a second figure, is also threefold: *hombres heterosexuales* are thought to be exclusively attracted to women; *bisexuales* attracted to women and men; and *homosexuales/gays* exclusively attracted to men "whether they or their partners are masculine or feminine in demeanor" (62). The last distinction is crucial, however, as the first model rendered impossible any relationship between straight-acting (masculine-identified) male partners.

More moving and fluid than these taxonomical grids are Carrillo's brief and suggestive sketches of his informants at the end of the book, where his descriptions range or flow over the fixed terms, intersecting unpredictably with age, class, and profession. Thus, the Antonio we already met is "a twenty five year old man who identifies as *gay*. He has worked since age nine and is now general administrator at a corporation. He lives alone in a rented apartment in a middle-class neighborhood" (Carrillo 2002, 312). Eduardo, on the other hand, is "a twenty seven year old man who identifies as *heterosexual* but who also has a sexual interest in men. He works and lives with his middle class family" (312). Elsewhere two college students identify themselves simply as *hombres* or *normales* (313), while a slightly older informant "knows the term *heterosexual* but does not apply it to himself" (313). The young woman, Martha, on the other hand, identifies as *femenina* and is unfamiliar with the term *heterosexual* (315). Clearly both sex/gender and object-choice models have some difficulty accounting for the messy and passionate lived experience and self-identification of subjects in Guadalajara.

Sergio de la Mora's *Cinemachismo: Masculinities and Sexuality in Mexican Film* (2006) might at first sight seem to have little in common with Carrillo's sociological study. Its cover (from *Los tres García* ["The Three Garcías," Ismael Rodríguez, 1946]) shows hypermasculine superstar Pedro Infante, arms defiantly folded, staring off into the distance as a bevy of black-clad beauties kneel supplicant at his feet. And in his introduction (titled "Macho Nation") de la Mora sets out his critical interest as: "The various strategies used by cultural producers and consumers to negotiate and contest cinematic representations of national identity that depict highly gendered and sexualized roles of men and women" (De la Mora 2006, 1). (Note the lexical similarity here with Carrillo's "strategies" and "roles"). To this effect de la Mora offers close analyses of what he identifies as "three traditional genres and a subgenre": "The revolutionary melodrama, the *cabaretera* (dance hall) prostitution melodrama, the musical-comedy 'buddy movie,' and the picaresque *fichera* brothel-cabaret comedy" (1).

Such cinematic genres both rely on and contribute to official discourses in postrevolutionary Mexico, most notably the machismo that is "the distinctive component of

Mexican national identity . . . rivaled only by the nation's deep religiosity manifest in the cult of Our Lady of Guadalupe, Mexico's patron saint" (De la Mora 2006, 2). De la Mora traces the history of such structuring archetypes throughout Mexican cinema, proceeding to parse one much-studied gay-themed film (Arturo Ripstein's *El lugar sin límites* ["Hell without Limits," 1978], 119–34) and even adding a postscript that pays loving attention to a more recent male pinup, Gael García Bernal (163–70).

Yet de la Mora begins (like Carrillo) with an anecdote: the "shock and delight" he experienced as a young gay man on coming across a nude photo of Pedro Infante in the shower (De la Mora 2006, viii). And de la Mora's argument throughout his book is that official binaries are much more slippery than they might at first appear. Thus, the publicity picture on the cover of the book reverses the traditional economy of narrative cinema by positing (like that illicit shower snapshot) the male as the object of visual pleasure: while his female acolytes are clad in all-enveloping black, the spectacularly handsome Infante sports the huge hat and elaborately embroidered vest and pants of the *charro*, or cowboy (an iconography born in the state of Jalisco, of which Guadalajara is the capital). Likewise, in the buddy comedies explored by de la Mora, homosociality (or companionship between men) is only blurrily distinguished from homosexuality (sexuality activity between them). Masculine privilege is, indeed, structurally reliant on the exclusion of subordinate women and effeminate men in order to secure its continuing dominance.

Moreover, as in Carrillo once more, nightlife is a privileged time and location for a sexual negotiation and expression that allows men (and even women) a certain relief from the rigors of daytime social intercourse. Beyond, then, both the gendered and sexual object choice paradigms cited by researchers in this field, de la Mora shows how, in film at least, identification and desire are fluid and fluctuating. Thus, the biologically male Manuela of *El lugar*, a transvestite in a brothel, acts and is acted upon as a female *cabaretera* or *fichera*, and is courted by an aggressively masculine *hombre normal* (De la Mora 2006, 123). Likewise, García Bernal and Diego Luna, the nominally straight boys of *Y tu mamá también* (Alfonso Cuarón, 2001), who spend the film pursuing an older woman, will, finally, make love with each other, exploring an erotic bond that they can neither name nor openly acknowledge (176–77).

To return to Hernández (who goes unmentioned in de la Mora's book), it is self-evident that he is fully aware of the sociological conditions of men who have sex with men in Mexico, however such men choose to define themselves. But, as we shall see, the queer auteur is (like Carrillo's case study, Antonio) traditional by nature, in spite of his love of novelty. Hernández will prove to be steeped in a cinematic history that suffuses his markedly modern features.

In the rest of this chapter I offer close readings of two contrasting feature films (*El cielo* and *Yo soy*), before turning to Hernández's lesser known shorts. I will suggest that *El cielo* shifts in focus from the sociological to the cinematic, while *Yo soy* moves in the opposite direction, from the filmic to the real. Hernández's shorts, meanwhile, occupy a perhaps more challenging position than his features, exploring fragments of time, space, and subjectivity and blurring together past, present, and future. What Hernández's films share, however, is the gesture of the extended hand with which I began this chapter: an amorous invitation to the character within the fiction that is redirected to the complicit art house or festival audience outside it.

From the Social to the Aesthetic: *El cielo dividido* ("Broken Sky," 2006)

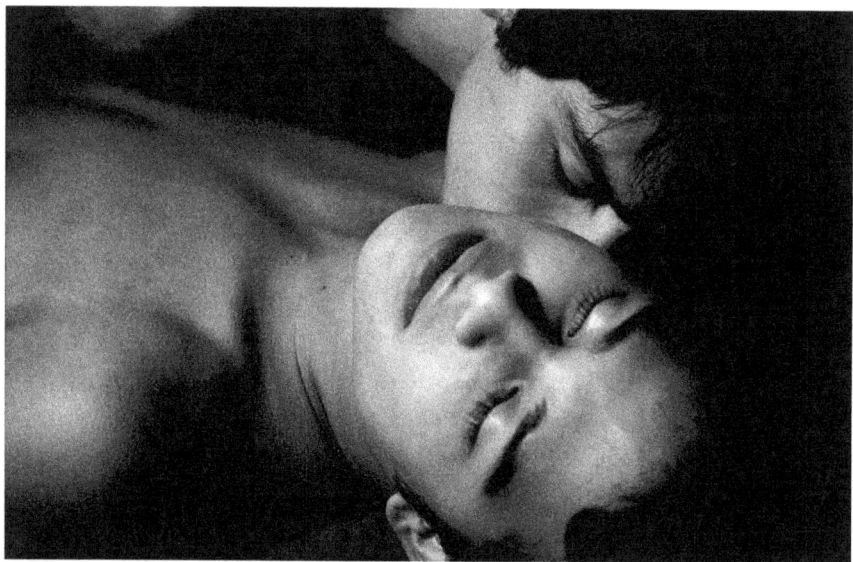

Gerardo es flechado y penetrado con sólo una mirada hasta lo más hondo de su alma y de su corazón por Jonás. Desde ese momento, comienza a manifestarse su amor y su lenguaje; un lenguaje que no sabe de palabras, un lenguaje espiritual y psíquico que culmina, en cada encuentro, con el fuego de la entrega mutua. Todo es perfecto hasta que la tentación invade a Jonás, quien vuelve a sentir ese flechazo; pero esta vez no con Gerardo. Jonás es separado de aquel que se convertirá en el amor de su vida. Desde ese momento, no piensa en otra cosa y comienza a rechazar a Gerardo, quien no se cansa de insistir en buscar los brazos y el cuerpo que siempre lo transportan a "otra dimensión," pero por mucho que intenta no lo consigue. Es entonces que Gerardo es cortejado por Sergio, quien, poco a poco, logra su entrega. Mientras, Jonás continúa soñando con su espejismo hasta que descubre la relación de Gerardo y también, el error que ha cometido. ¿Será demasiado tarde?, ¿Se destruirá para siempre su unión espiritual? ("Synopsis of *El cielo dividido*" 2006)

Geraldo is pierced and penetrated to the depths of his soul and heart with just one look from Jonás. From that instant his love and language start to make themselves felt, a spiritual and psychic language which culminates, with each encounter, with the fire of a mutual surrender. Everything is perfect until temptation overwhelms Jonás, who feels the piercing look of love once more, but this time not from Gerardo. Jonás is separated from the man who will become the love of this life. From that moment he can think only of one thing and starts to reject Gerardo, who tirelessly insists on seeking the arms and body that always transport him to "another dimension," but he does not succeed however much he tries. It is then that Gerardo is courted by Sergio, who, little by little, makes him surrender. Meanwhile, Jonás carries on dreaming of his fantasy man until he discovers Gerardo's new relationship and also the mistake he has made. Is it too late? Will their spiritual union be forever destroyed?

The official synopsis of *El cielo*, with its talk of "spiritual language" and "another dimension" is noticeably abstract. Yet in the first sequence the mobile camera follows protagonist Gerardo (played by Miguel Ángel Hoppe, who, like the rest of cast, would never work in film again) as he walks alone through a concrete location: the vast Ciudad Universitaria, or campus of the UNAM (National Autonomous University). He even passes in front of the famous mural on the Central Library wall, a specific site that no Mexican could fail to recognize.

Two scholars of the film recreate this conflict, focusing on the objective and the subjective, respectively. Thus, Gustavo Subero is centered mainly (like Carrillo) on the sociology of everyday gay life, exploring such material issues as race and class (Subero 2013, 96–127). Bringing the auteur down to earth, he even compares Hernández's oeuvre to that of the porn producer I examined in the first chapter: "The work of Mecos Films, similar to that of Julián Hernández, is the first real attempt to visually acknowledge ethnic syncretism within Mexican culture, specifically queer culture, by portraying characters and situations that are not ethnically coded" (187).

Subero's chapter on *El cielo* in his book *Queer Masculinities in Latin American Cinema* (2013) is titled "Where Gay Meets Race" (91) and celebrates the director's exploration of "Indo-mestizo bodies" (97). Subero also places *El cielo*'s triangle of lovers within distinct "socio-economic backgrounds [that] can be easily recognized through the mise en scene of their homes" (113). Thus, dark-haired Gerardo lives with his mother in a slightly run down but comfortable project, while lighter-complexioned Jonás appears to lodge on his own in a more tony, minimalist apartment. The real-life university campus where most of the action takes place is presented as an "idyllic, accepting, and tolerant environment" to the young lovers who study there (Hernández is well aware that this is somewhat idealized) (114–15). And in spite of the film's habitual lack of dialogue, *El cielo* also offers some evidence of social support networks for its characters: Gerardo's mother is quietly content for her son to bring his new partner home for the night and later consoles him when they split up; mestizo-featured Sergio, who finally woos and wins Gerardo, is supported in his own troubled love life by a female friend.

According to Subero, fashion is vital here too as a system of social inscription. Sergio's lip piercing is read as a gesture of "fictive kinship" with a new gay community (Subero 2013, 121). The other boys' trendy T-shirts are held to signal a gay "subculture" that "permits members . . . to identify and recognize each other" (125). Yet this vestimentary code is slippery. Subero writes that Gerardo's "McQueen" T-shirt references the British designer Alexander, proof of the character's literacy in the transnational language of fashion (125). In fact, it shows the face of Steve McQueen, the stylishly macho star of US film in the 1960s.

Subero himself makes another cinematic connection here. The Gerardo who, he infers with no explicit evidence from the film, is caught between white and brown, upper and lower class, also serves as "the stereotypical heroine of the Mexican melodrama who is torn between two male figures who will go to any lengths to win her love" (Subero 2013, 117) (such situations are characteristic of the Golden Age buddy movies treated by de la Mora). Subero further suggests that "this type of representation of the queer male protagonist [is] quintessentially feminine . . . not because he behaves in a [*sic*] unmanly or feminine manner but because he does not exercise or impose power over the people around him" (117). Supposedly comparable to the Virgin Mary in his self-denial, Sergio is also "arguably depicted as a *chingado* [easy screw], because he allows the other two males to exercise power over him and constrain his own desires" (118). By the end of the film, however, "he will break with the cycle of the drama of identification that depicts *Malinche* as both a national whore and a victimized figure" (119). Both Virgin and prostitute, Gerardo is said to be suffused with the tragic ethos of Mexican cinema of the Golden Age, even as he is enmeshed in contemporary expressions of race, class, and subculture which that cinema could not have anticipated.

Yet the limits of Subero's sociological reading (already qualified by the flows of filmic tradition) stand revealed when confronted by the text of the film. After a first half hour of romantic idyll, Gerardo will indeed spend much melancholic screen time mooning over the Jonás, who has rejected him for another man. This third man (the "fantasy" of the synopsis) was glimpsed in a pulsating nightclub, a sensuous space of pure light and color where bodies and subjects are abstracted (as Carrillo noted) from the rigor of daytime rules. But the erring Jonás will later come to occupy Gerardo's position as the weeping, spurned lover, when, now regretful, he spies his ex with a new partner. And, in the most explicit reference to Golden Age cinema, it is not Gerardo but Jonás who will leaf tearfully through an illustrated volume on classic director Emilio "El Indio" Fernández written by renowned film historian Emilio García Riera. It is in this cinematically loaded location that Jonás has placed treasured snapshots of his lost lover.

In terms of plot positioning and connection with filmic melodrama, then the (supposedly, relatively) upper-class, lighter-skinned Jonás is at least as feminine-identified as his lower-class, browner lover. And unlike Subero, Hernández would seem to have little interest in social roles or strategies. In *El cielo* (as in Carrillo's case studies) the need for love is total and (initially at least) mutual. And Hernández's characters are rigorously deprived of those cultural clues (age, class, profession) that inform Carrillo's biographical sketches.

Film technique also diminishes these social markers. Skin of all complexions is shot lovingly in extreme close-up, heightening our sense of a tactility that eludes or elides the social meanings of pigmentation. And explicitly rejecting the gender-based paradigm of sexual identity (set out Carrillo's first figure), which Subero seems to be citing here, Hernández's lovers cannot be reduced to *activos* and *pasivos*, as they are evidently *internacionales*. Thus, in the second sequence of the film (after Gerardo has wandered through the campus) we are offered a three-minute, unbroken take of the protagonist's first lovemaking with Jonás. It is a flip/flop fuck (relatively unexplicit, however) in which each partner takes rapid turns in each sexual role. Although the two actors are clearly faking the sex acts, the absolute need for love of their characters (like that of Carrillo's case study Antonio) evidently permits of no passion-killing delay to negotiate condom use.

Likewise, when Gerardo first has sex with the older, darker, and more working class Sergio, it is, surprisingly perhaps, the former who penetrates the latter. Moreover, innocent of labels and, indeed, hostile to language itself, Hernández's silently brooding and cruising boys never feel the need to define themselves, enjoying a same-sex eroticism that is taken for granted by all (even heterosexual observers on campus look on fondly at displays of affection between the men). In the only classroom scene we are shown, Gerardo listens as the professor describes Plato's theory in the *Symposium* of the splitting of primal beings, which caused humans to seek ever after their perfect other half. Plato's parable is an origin story for homo- and heterosexual love alike.

The authentic locations, whose role in social realist cinema is to constitute a determining environment for the characters who inhabit them, are in *El cielo* abstracted into pure film, pure aesthetics. It is true that sometimes these places are recognizable:

Jonás spies on his ex kissing Sergio at the base of the Angel monument on Reforma (like the kids in the webseries *Al final del arcoíris* a couple of years later). But a less aesthetically pleasing monument, the hulking sunken circular plaza of the Insurgentes subway station, is lit in such a way (thick black shadows, intensely colored lights) as to remake its grimy concrete into a kind of stylish abstract expressionism. The film's original soundtrack by Arturo Villela, with its classical harpsichord, cello, and oboe, also tends to abstract the action, to remove it from the messy particularity of everyday life.

A sequence on the footbridge across Insurgentes once more, farther south at the university campus, is shot in tight close-up, thus isolating it from its actual context and everyday use by students. And when the boys come back to Gerardo's lower-middle-class housing project at night, city lights glow red, green, and yellow behind them as they kiss on the walkway outside the apartment, a neon composition of multi-hued delight. In perhaps the most striking scene of the film, the university's Olympic swimming pool is first shown distantly blue behind the male couple as they embrace in close-up on the bleachers. Then, with a jarring cut of 180 degrees, the pair seems suddenly tiny, lost in the vast horizontal geography of the sporting structure. Location and technique have not a social but a narrative effect here, signaling the ongoing split between the two lovers.

In "Landscapes of Subjectivity in Contemporary Mexican Cinema," an article that makes extended reference to *El cielo* as well as films by other young directors, Laura Podalsky (2011) also cites this sequence, reproducing a geometrically patterned still with the caption "The isolation of any-space-whatsoever" (173). The caption is in keeping with Podalsky's general argument, according to which "through formal experimentations with spatial and temporal markers, the films rewire the relationship between the viewer and pro-filmic events and, in the process, position the film-makers themselves as auteurs on the global art scene" (161).

For Podalsky, who is less interested in sexuality than Subero, this filmmaking still has a vital gendered element:

> Such films articulate a new understanding of masculinity within the context of Mexico—one that is less interested in tying man to nation or in representing man as integral, impenetrable subject. Despite occasional nods to Mexican history, the films do not situate their protagonists as representative of particular generations or as symbols of Mexican manhood in general, as did works from the so-called Golden Age. (162)

In contrast to this tradition, the new films (and Podalsky includes with Hernández such festival favorites as Carlos Reygadas and Fernando Eimbcke) avoid "references to particular socio-historical contexts in order to bring into relief micro-tales of male subjectivities in transition" (Podalsky 2011, 162). Time and space are also remade, remodeled in similar, fluctuating fashion:

> Space does not function as a backdrop for the unfolding of dramatic conflicts or as an allegory delineating national identity. Instead rural and urban landscapes function as the material register of male subjectivity. Time is not registered in terms of history, but rather through movement. (163)

More specifically on *El cielo*, Podalsky writes that its "urban settings serve as the site of subjective unfolding" (172). The "geometric compositions," even when based on identifiable locations, signal in a sexual context materializations of "the symmetries and asymmetries of homosexual desire" as they slowly come into focus. Or again, "shot in particular ways, the urban landscape makes visible the rhythms and texture of [young people's] experiences that are only beginning to now crystallize" (172). Part of this emergence, perhaps, is the fact that Hernández's young men are neither effeminate nor aggressively masculine, having transcended the gender-based paradigm still acted out by so many of Carrillo's real-life informants and de la Mora's cinematic archetypes.

While Podalsky's auteurist and formalist perspective (based in part on Gilles Deleuze) seems far from Subero's social and political approach, she ends her article by making explicit a connection with the material conditions in which her films are produced. Thus, the rise of private investment in film through corporate tax credits (which I mentioned earlier in the context of the Guadalajara market) might be related to these films' increasing reluctance to address a Mexican national identity that was formerly fostered by funding from government agencies (Podalsky 2011, 173). And Podalsky cites, like myself, the sociologist Carrillo as part of an argument for the existence of a recent qualified challenge to male privilege (from women and gay men) that has led to the emergence of "a more plural understanding of what it means to be a man in Mexico" (175). Just as Subero's objective criteria led him on to cite subjective fantasies fostered by Golden Age film, so Podalsky's subjective rhythms and textures invite her back to acknowledge objective indexes of change in the film industry and in society.

Internet sources also mutate the nature and audience of a film that is now almost a decade old and still in a state of transition. In February 2016 Hernández posted a link on Twitter to deleted scenes on Vimeo (including a more explicit shower sequence by Fernando Arroyo, the actor playing Jonás) (Hernández 2016). These served to extend a filmic text that at 140 slow-paced minutes is already relatively long. Meanwhile, an unusually detailed blog by *El cielo*'s script supervisor offers a parallel narrative to the film itself (Pérez Mancilla 2007a). Offering unique insight into the production process, the blog complements both subjective and objective aspects of the film. Thus, we are told of the many material difficulties of filming on location with a tiny budget (at one point the cast and crew were abused on the street as "maricones"). But the blog also testifies to the emotional dramas behind the camera: the auteur's pervasive melancholy and the cruelty with which he sometimes treated his inexperienced cast in an attempt to produce the desired artistic effect. Fernando Arroyo (Jonás) is, once more, singled out for ridicule as a "homophobe" for his reluctance to perform in the gay love scenes.

As we have seen, *El cielo* is set in a university environment where young men's sexual roles and strategies are not explicitly cinematic, even though they conceal real-life dramas on set. We turn now, however, to a more recent feature by Hernández, which Subero and Podalsky were unable to treat, in which a director's problematic, even abusive, relationship with his actors takes center stage.

From Film to Life: *Yo soy la felicidad de este mundo* ("I Am Happiness on Earth," 2014)

Emiliano, desde su mundo de director de cine, explora sus procesos y trata de conectar con su realidad inmediata, la historia que se filma se mezcla sin miramientos con su realidad cotidiana, su mundo real parece estar visto siempre por el lente de la cámara y su realidad está emplazada entre un punto de vista que resuelve sus conflictos morales en las acciones cotidianas; confundido, solo, siempre frente a la pantalla que es su realidad transfigurada, su realidad mesurable, controlable, manejable, escucha solitario esa canción que se repite como una oración que te obliga a seguir el intento de amar. ("Yo soy la felicidad de este mundo" 2014) (punctuation as in original)

Emiliano looks at his life with the eyes of a film director, mixing the objective reality with the processes of the artistic creation. The story he is filming flounders with his daily life, until his world is trapped in the lens of his camera. Confused, always alone and in front of a screen, now become a transfigured reality, but at the same time a measurable, controllable and manipulable one, he listens in loop to a song: one of those songs you sing or repeat as a prayer and forcing you to remember, believe and convince yourself.

In spite of the repeated stress in the Spanish-language synopsis above (the English is slightly different) on a reality that is "immediate" or "everyday," *Yo soy* is surely Hernández's most abstracted and hermetic feature to date. Its plot structure is radically fragmented, divided into three unequal and barely connected parts. In the first and third, the protagonist (Emiliano, a film director played by telenovela heartthrob Hugo Catalán) initiates somewhat morose and inconclusive affairs with a dancer and a male sex worker, respectively; while in the second we are shown a portion of the film within the film: a modern dance piece that includes explicit scenes of masturbation.

Surprisingly, perhaps, after its premiere at the Morelia festival, *Yo soy* achieved the rare opportunity of a theatrical release, albeit short and very restricted, in the United States. It was then immediately available there for streaming on Amazon Instant Video (it remains accessible on Kanopy, an online art movie streaming service). Appealing to new windows of distribution, Hernández (or Fiesco) thus hoped to avoid the distribution conundrums that have confounded Mexican independent cinema until now and were frequently commented on in the trade press over the course of their joint career.

The defiantly minority aesthetics of the film are shown in several ways, both new and old. Thus, *Yo soy* benefits once more from Alejandro Cantú's luminous photography (it was shot on the now archaic or artisanal medium of 35 mm). Now, more than ever, Cantú uses intense light and dense shadow to transform male faces and bodies into sculptural compositions. The film also relies more heavily than before on an original soundtrack by Arturo Villela (who, as mentioned, also scored *El cielo*) which is virtuosically varied: pseudo-classical pieces are heard over the main two love-cum-sex stories, while the central dance or physical theater piece plays against an aggressively discordant modern soundscape. It remains the case that Hernández's preferred Mexican balladeer José José plays a central role here, mentioned even in the synopsis: his song "Dos" ("Two"), a hymn to the couple and to a true love that vanquishes all obstacles, plays with solemn irony (ironic solemnity?) at two vital points in this decidedly unromantic film. But Hernández also stages an unaccustomed ménage à trois to a plaintive and enigmatic recent song by Vetusta Morla, the

Spanish indie rockers who are known for their minimalist videos. Acting as a kind of virtual DJ, then, the director engages with new forms of distinction, enlisting new kinds of cultural prestige.

The cast, moreover (with the main exception of professional dancer Alan Ramírez), boasts some accomplished actors. The classically handsome Hugo Catalán, as the self-absorbed filmmaker antihero, has many credits in television and mainstream film (I treat his gay drama *Cuatro lunas* in chapter 4). The defiantly ugly Gabino Rodríguez, who has a wordless but physically challenging role in the central part of the triptych, is the long-time collaborator of austere festival favorite Nicolás Pereda, as well as a respected figure in Mexico City theater. Andrea Portal, who plays Sunny (a singer at a party in the first section, an aroused participant in the erotic second part) is the protagonist of lesbian dramedy *Todo el mundo*, which I also examine in chapter 4. Such more experienced actors offer a possibility at least of greater emotional expression on screen than the cute but blank young amateurs of *El cielo*.

And for the first time in his feature oeuvre, Hernández introduces a new theme: professional dance. The film opens (rather like Almodóvar's *Hable con ella* ["Talk to Her"], 2002) with a dance performance by a mature woman. Here veteran Gloria Contreras even looks passably like Almodóvar's Pina Bausch. Contreras is also glimpsed later in a TV interview where she speaks of her rejection by the cultural establishment in Mexico and her quest for true, real expression beyond artistic performance. These are sentiments that Hernández no doubt shares. What is vital in his film, however, is that dance, in spite of its importance here, is not directly representational or narrative. And *Yo soy*'s climax will prove to be a lengthy expert public performance by Octavio. Lit by harsh chiaroscuro spots and clad only in magenta shorts, he is fully abstracted from the everyday reality that Emiliano, Contreras, and the film claim to be seeking. This climactic dance number also notably fails to resolve the tragic love triangle on which the film's plot appears to be founded.

Also for the first time in *Yo soy*, Hernández's perfectionist aesthetic abstraction is openly self-reflexive, incorporating as it does into the fictional world a dialogue on the nature of art cinema and its real-life investments. Thus, Hernández's protagonist Emiliano is making a documentary on dance, because (or so says one dancer with some malice) "he likes dancers." The same dancer also hints slyly to Octavio that the director is looking for a star for his next film. Octavio's modest studio apartment, to which he and Emiliano soon retreat to make love, boasts a poster of Fassbinder's tragic drama *Veronika Voss* (1982) behind the bed, surely an unlikely choice for a youthful dancer of apparently humble origin (mestizolooking, the lighting consistently highlights his handsome cheekbones). More plausibly, Emiliano's more stylish apartment with its flat screen TV (Octavio's is an old chunky model) features Mexican movie paraphernalia, including posters for his own (fictional) films.

When Emiliano later calls in the services of blond boyish sex worker Jazén (newcomer Emilio von Sternerfels), the stage is set for the most explicit discussion about film in Hernández's work. Jazén is, implausibly once more, highly impressed by Emiliano's curriculum, exclaiming "Everybody saw that film!" on seeing the poster for Emiliano's opus on the wall. Jazén also defends "art cinema" against those who complain that "the characters don't speak and the films are too long" (frequent criticisms made, of course, of Hernández's own works). Hernández even incorporates tiny self-references or Easter eggs as winks to his long-term fans, nurtured by social media. Or perhaps they are intended for critics like me, who are keen to display their expertise in the details of the auteur's oeuvre. For example, when Emiliano, neglecting the Jazén who has now moved in with him and abandoned his former profession, is marking up a script, we see in a close-up of the pages that it is for Hernández's earlier feature *El cielo*. Or, again, when Jazén suggests a subject for a film (an older man takes a younger mute boy to a hotel), which is brusquely rejected by Emiliano, it is actually the premise of *David* (2005), a tender short directed by Hernández's producer Roberto Fiesco.

As can be confirmed by my interview with Hernández at the end of this book, the director identifies the character of rent boy Jazén with the female prostitutes (*cabareteras* or *ficheras*) of the Mexican cinema of the Golden Age, tragic heroines of musical melodramas. And the frequent breaks for performance in *Yo soy* (both musical and balletic) are reminiscent of the old-style numbers of such films in that they reduce the narrative to a willful standstill. The dance numbers also serve their traditional role as vehicles for an ecstatic release from the everyday that takes place in the hallowed space and time of nightlife (a sequence at a private party looks more like it was staged at a club). Yet, as Hernández also mentions in the interview included in this book, the sex worker character in his film is, unlike his cinematic precedents, not self-sacrificial. Jazén will bring an even younger, handsome client back to Emiliano's home, refusing payment for his services, and will finally walk out on the drug-addled director. Moreover, in the interview he gave me Hernández claimed himself, teasingly perhaps, to identify in his film not with the filmmaker but with the sex worker.

Yo soy thus confirms the long shadow of Golden Age archetypes, albeit transformed by their queering, so closely documented by de la Mora. And Hernández's characters might perhaps be read (reread) in the light of Carrillo's case studies also. But, even more than in *El cielo*, socioeconomic context is kept from us. All we know of Emiliano is that he a twenty-something film director who lives on his own, sometimes has sex with male prostitutes, and occasionally has a live-in lover. Dancer

Octavio, somewhat younger, also lives on his own and seems to be seeking long-term relationships with men. But in one unexpected scene he shares a threesome with two sexually assertive women (the sequence is set to the plangent Vetusta Morla number). As there is no discussion of sexual roles or identities in the film, however, it would be hard to label Octavio as "bisexual" using Carrillo's socially constituted object choice model. Similarly, it comes as a shock when sex worker Jazén tells Emiliano at one point that he needs to leave "for a ten o'clock class." We had no idea that he was a student. The three participants (actors? dancers?) in the central physical theater piece are silent, nameless bodies, illuminated only by a poetic voiceover as abstract as their movements.

Yet if we reread these central themes of the film (cinema, sex work, dance) in a new light we find a way through or out toward the "real life" apparently sought by a character and director so mesmerized by the lens. Hernández lays bare the device of his art cinema and its relationship to homosexual desire. An early scene shows the (fictional) crew led by the (fictional) director shooting one of the (real-life) director's signature 360-degree pans. For the first time, we see the circular tracks that enclose the performer. The ubiquitous scripts and screens seen on screen not only suggest a hermetically narcissistic life within the film (Emiliano watches himself making love on that flat screen TV) but also the material reality of cinema as institution outside it. On location Emiliano asks his director of photography how much time they have left to shoot (they are filming a dance piece in an urban wasteland). At home, he wonders if he has enough money to pay a prostitute for extended hours. Later he scrolls through profiles on an escort site or negotiates a modest fee of 800 pesos or about fifty dollars ("all included") with a male sex worker in a park. Cinema and sex work are intimately integrated exercises in Hernández's management of space and time. To return to Carrillo's terms, Emiliano's cultural "scripts" and "strategies" are at once sexual, cultural, and financial.

Hollywood Reporter's negative review of a film it calls "beautifully photographed but tedious," notes that "gay audiences no longer rush out to see LGBT films even if they have explicit sex" (Farber 2014). Yet, aware perhaps of trends in international auteur film (or indeed his own shorts), Hernández includes in *Yo soy* for the first time sex scenes that are real, not mimed like those in *El cielo*. In the central section of the film (which lasts around thirty minutes) we see moments of male and female masturbation that intersperse a performance of physical theater whose everyday movements (slumping against a wall, writhing on a floor) and setting (an unexceptional home) are, like the erect penis displayed for our attention, disconcertingly close to real life.

Blurring the boundary between art movies and pornography (as Subero suggested of Hernández's love of naked mestizo bodies), Hernández thus incorporates a trace of the real into his world of airless abstraction. Indeed, the fact that he shoots still on 35 mm means that that trace is tactile, heir to the ontology of the photographic image famously invoked by André Bazin so long ago. That ontology is also enmeshed with the messy materiality of finance: Emiliano plans a visit to the ATM to pay an escort, the film's credits thank rather FOPROCINE, the state-organized capital fund for developing quality film. This sense of financial risk renders Hernández and Emiliano's continuing cinematic enterprise more precarious and perhaps more moving.

As mentioned earlier, the film ends with a climactic, classically inflected performance by Alan Ramírez, a member of Mexico's National Ballet. It is established that this scene takes place at the Sala Miguel Covarrubias, the dance auditorium of Hernández's beloved UNAM (it is one of the few recognizable urban locations in the film). The specificity of the setting reveals how dance, however abstract it might first appear, is always eventually enmeshed (like cinema) in institutions and is inevitably embedded in society and history. Yet the principal intertext of *Yo soy* remains José José's song "Dos" ("Two, we are two / only two to feel a true love in the heart"). After having sex that is shown only later in a tape on the TV screen, Emiliano and Jazén sit on the bed side by side in two-shot singing along to this romantic ballad in praise of the couple. Earlier on, however, we have twice heard Emiliano in a television interview attacking gay marriage as "obsolete."

The very last shot of *Yo soy*, contradicting the film's optimistic title, is of the protagonist on his own, rejected by both lovers, weeping as he listens and sings along once more to the same song. Hernández's rigorous omission of social context for his characters renders our identification with them problematic. And his frequent self-conscious ironies make affect difficult to accede to also (I for one, and unlike Emiliano, did not cry on repeated viewings of the film). Yet *Yo soy*, supposedly too much the same for Mexican critics, marks a difference in Hernández's feature oeuvre with its close exploration of female bodies as well as male and its amorous attention to dance. And, fighting its way from the cinematic to the social, the film makes a gesture perhaps in spite of itself toward a concrete reality that can be approached only through rigorous aesthetic form. This is perhaps the final meaning of Emiliano's repeated arm and leg extensions in his dance piece, reaching out to the audience (on screen, off screen) even as he breaks up definitively with his former, feckless lover.

In Praise of the Fragment: Three Shorts (2007, 2010, 2015)

In 2015 one of Roberto Fiesco's shorts won an award. Interviewed by Leftist daily *La Jornada*, he is quoted as saying that the short is "free territory . . . the bastard brother of [feature] film" (Cruz Bárcenas 2015). Although Fiesco laments shorts' lack of distribution, saying directors need to seek out windows beyond theaters and festivals in television and the Internet, he also claims that shorts tend to be seen by industry insiders, "film people." There is thus a pragmatic, as well as an artistic, motive for making them.

It would not be surprising if Fiesco's partner Julián Hernández were to share such views about shorts. Certainly, Hernández has throughout his career continued to practice the genre, making some twenty titles, in tandem with the fiction features that Fiesco has yet to direct. In this final section of my chapter, I address three representative shorts made by Hernández over a decade, which, I argue, deserve the critical appreciation that has previously been restricted to his four features. Moreover, in Hernández's practice at least, the genre of the short, in spite of its limited distribution, poses a challenge to the prestige and indeed the integrity of the feature film as object and the homosexual as emergent subject.

The first film under consideration is *Bramadero* (2007). This short is cited within Hernández's third feature, *Rabioso sol*, in a scene set in a ruinous porn movie theater, where the title appears on a poster and some footage can be glimpsed, blurrily, on screen in a darkened auditorium frequented by shadowy cruisers. Barely incorporated into the body of the auteur's features, *Bramadero* was exhibited also in a somewhat marginal manner at the Morelia festival. There it shared a midnight bill with historic silent pornographic titles, restored by the Filmoteca, the cinema archive of the UNAM (Cárdenas Ochoa 2008). Anticipating Fiesco's words on the short in a more restricted context, the journalist writes of the screening that it converted the festival into a "free territory for filmic expression."

The unusual title of Hernández's film suggests the brute instinct of nature, translating as it does as "rutting ground," the place where wild animals come to mate (a dictionary definition of the rare Spanish word is offered in the final credits). And in this urban location (a high-rise building that is either being built or demolished with ambient street noise heard throughout), two men come together to brood, pace, fuck, and fight. Silent as they are (there is no dialogue), the handsome figures, frequently unclothed, suggest that absolute need for love, without negotiation or qualification, shared by Carrillo's informants and Hernández's characters elsewhere. (Here, however, the two men do use condoms for anal sex.) This extremity of passion can culminate only in death, when one partner strangles the other. It is a vestigial, melodramatic love story that, in spite of its transgressive explicitness, owes something to the emotional excesses of Mexican cinema of the Golden Age. And the shooting style, with its trademark 360-degree pans, is as stylized as anywhere in Hernández's oeuvre.

However, the urban landscape, glimpsed behind the figures throughout the short as they silently quarrel or make love, is of the essence here. This is not just an authentic location; it is a positively dangerous one. The cast (and crew) are perilously placed high above an urban vista on an unmade concrete floor that has no external walls. The anonymous space is *in* the city, but not *of* it. Perhaps it is the remains of the soon to be derelict hotel in which *El cielo*'s lengthy opening sex scene was shot? That building was subsequently stripped down and converted to condos as part of the lengthy process of gentrification of the degraded Historic Center. Likewise, the Cine Teresa, whose sign is glimpsed at one point in *Bramadero* behind the couple as they pace their urban aerie, was a historic theater at that time programing porn movies. It closed in 2010 to reopen as the Historic Center's outpost of the Cineteca, the national film institute ("Cine Teresa" 2013). Mexico City's urban space inevitably carries a historicity within it, bearing the trace of a lived experience in its look and sound for both the director and his audience.

There is a stubborn particularity embedded in the city's inhabitants, even here, in this would-be elemental rutting ground. As so often in Hernández, even if both young men are equally inexpressive, one is darker, more mestizo-featured, another fairer, more

Mediterranean. Moreover, the gym-trained body of the latter, lighter-skinned one is recognizable to connoisseurs of Mexican porn as belonging to "Tomás" in Mecos's first *Selección mexicana* audition movie. Here in *Bramadero*, where he has a different name, he is handsomely stone-faced as he submits to the intense gaze of his partner and, no doubt, that of the avid cinephile and other publics. *Bramadero* is currently posted at full length on the porn site XVideos, where it had received over 350,000 views, surely Hernández's biggest audience by far.

In *Selección*, however (as mentioned in chapter 1), "Tomás" is smiling and engaging, even as he abandons himself with evident relish to a variety of "international" sex acts with his Mecos partner, the animalistically named (but also sexually versatile) El Puma. And in interview before his scene, "Tomás" offers director El Diablo a backstory reminiscent of Carrillo's case studies of emergent subjectivity. He says he was initiated into sex with penetration by a male relative at age eleven. Sadly, the other actor in *Bramadero*, murdered in this short fiction, was soon to die also in real life (Pérez Mancilla 2007b). Even the most stylized cinema aesthetics, then (*La Jornada* praises Hernández's use of the sequence shot as well as his reflections on "homosexual problematics"), is no refuge for its participants from the irreducible dangers of the real.

My second short seems quite distinct. *Atmósfera* ("Atmosphere") is Hernández's segment from a portmanteau movie made by eight directors called *Sucedió en un día* ("It Happened in a Day," 2010). All the segments were shot far from Hernández's favored Mexico City, in Playa del Carmen in Yucatán, the beach resort that is an upmarket rival to the mass tourist destination of Cancún. And each segment was filmed in just twenty-four hours, in accordance with the project's title.

Unlike *Bramadero*'s gritty contemporary realism (albeit located in a set suspended high above the city streets), *Atmósfera* reads like science fiction, a genre barely essayed in Hernández's features (*Rabioso sol* does boast fantasy sequences shot among pre-Hispanic monuments). Filmed, unlike *Bramadero* again, in glossy black and white, *Atmósfera* seems to depict a future dystopia in which an ecological catastrophe makes it impossible to venture outside. The enigmatic location is an abandoned hotel where three figures coincide. A young woman leaves a lipstick trace on a window. A young man, flitting on a skate board, traverses the hotel corridors. And, offering fetishistic focus on another sculptured male body, respected young actor Harold Torres does push-ups or ventures in outsize goggles to the swimming pool. Finally, however, the three characters, apparently ignoring or eluding deadly atmospheric peril, escape out to the beach where they strip off and bathe in the ocean.

As in the dance sequence of *Yo soy*, the appeal to a threesome (two guys and a girl) once more problematizes the idea of the couple, most especially the homosexual couple that was critiqued also in *Yo soy*. But the theme of environmentalism, even when treated in Hernández's signature style, seems very new. And the suggestion that nature can heal the wounds of ecological distress seems far from the celebration of famously polluted Mexico City and its still grimy Historic Center, a cinematic love affair to which Hernández (and Fiesco) remain so deeply attached. *Atmósfera* thus expands the auteur's thematics even as it holds close to his familiar aesthetics.

Close analysis of this short's final five-minute sequence, filmed on location on a Yucatán beach, confirms the connection between film technique and sexual and social issues. The scene begins (as it ends) with an idyllic landscape shot of unpopulated white sand and turquoise ocean, strongly contrasting with the monochrome location of the abandoned hotel, which is all we have seen until now. The girl, clad in a hot pink minidress and with a large camera around her neck, comes into shot, gazing out into the ocean. Next the first boy looms smiling into the camera on the left of the frame and the second boy moves in on the right in the following shot. Both are carrying skateboards. In a traditional example of the eyeline match familiar from continuity editing, each boy looks left and right, respectively, establishing their unspoken connection with each other (there is no dialogue in the sequence). Finally, all three characters appear in the frame and the camera pans left and right once more, following the girl's gaze, as she connects with her male companions. There is thus a strict symmetry of looks (from both camera and characters) suggesting a sexual or social synthesis between these handsome but enigmatic figures.

A song now comes in on the soundtrack. Appropriately enough, it is "La playa" ("The Beach") by 1960s songstress Monna Bell. Each actor then strips off and they sport together in long shot in the sea, while the girl's camera is prominent in the foreground, abandoned on the sand. Next, a high angle shows the naked threesome arranging three photos on a skateboard. Within a single, extended take, the camera tilts up to follow the characters as, silhouetted by the dazzling sun, they return to the ocean, tilts down to the photos once more, and then back up to the sea, which is now empty of humanity. It is a technique playing with presence and absence that is very similar to the one we have seen in *El cielo dividido*, whose boys seem always to be disappearing from their beds as the camera roams rooms. Meanwhile, Monna Bell's smoothly seductive song reaches its climax with an appeal to "the immensity of our love."

Atmósfera seems to suggest a healing of ecological stress through baptism in an idyllic nature and, perhaps, an erasure of sexual conflict through a bisexual utopia embodied in an equally idyllic threesome. But through his rigorously symmetrical framing and camera movement, Hernández characteristically reminds us of the abiding artifice of cinema. Indeed, the girl's camera remains on the beach, even as the people it has shot have disappeared into the immensity of the ocean with their silently, cinematically affirmed love.

In my third and final short, the award-winning *Muchacho en la barra se masturba con rabia y osadía* ("Young Man at the Bar Masturbating with Rage and Nerve," 2015), Hernández returns to his beloved city but with a radically new departure and genre (the film is not as graphic as its title suggests). Cristhian [*sic*] Rodríguez, a professional dancer and sex worker, is the protagonist of this documentary, which is, in turn, moving and exciting.

The subjects in Hernández's fictions are, as we have seen, typically stripped of the particularity of economic determinants stressed by sociologists. But here in documentary, by generously lending a voice to Rodríguez, Hernández facilitates a valuable and enjoyable microhistory of gay life in Mexico. It is one analogous to those of the informants in Carrillo's book, but now enriched by the audiovisual dimension. Cristhian, an immigrant to the capital from a village in provincial Mazatlán, tells his story of a rigorous training in dance, a creative vocation that later required him to undertake sex work to pay the bills. The observant viewer will note that, although now slightly older and boasting a butch buzz cut, the resourceful Cristhian is in fact the same person as *Selección*'s smiling "Tomás" and *Bramadero*'s sullen lover-cum-murderer.

Shot at home, play, work, and in the steam baths, Cristhian proves as engaging and seductive as in his interview for Mecos eight years earlier. And he has still maintained the discreetly muscled physical type preferred by Hernández. The linked themes of dance and sex work appear entwined once more, as twin disciplines of the body or commitments of the individual to a cultural or sexual collective. Yet while in the fiction *Yo soy* the two roles were distributed among two characters (the director's lovers), here in the documentary they are more problematically combined in one. However, if Cristhian might initially be seen as a modern *cabaretera* or *fichera* (de la Mora's tragic Golden Age heroine) he is in fact as self-assured in the daytime as he is in the protective obscurity of the night, when he tends to exercise his two professions.

From a story that is strictly self-related (we hear no questions from the interviewer) there emerges a new theme in Hernández's oeuvre: effeminacy and the perilous emergence of a gay identity in a working-class setting. Cristhian, now masculine in demeanor and sexually versatile, confesses that originally his only conception of homosexuality was as wanting to be a girl. And, as a victim of small-town homophobia before coming to Mexico City, he could not imagine how two *maricones* could have a relationship together, a possibility in which he subsequently took such serial delight. Cristhian's story unselfconsciously charts, then, that fluid transition between the gendered and the object choice models of homosexuality that are so drily set out in Carrillo's figures. Born a *maricón*, Christhian is now a full-fledged *gay*.

We saw that social markers are rigorously kept off screen in Hernández's fiction features and can be inferred only (as Subero does) by elements of the mise-en-scène. But those life-making factors of class, region, and profession make a moving entrance here in the documentary short. It helps that while Hernández's feature protagonists are largely silent, the star of his short is so voluble. And allied to this emergent subjectivity, there arise (as Podalsky suggested) new landscapes of subjectivity, both interior and exterior. As a choreography teacher, Cristhian leads his pupils in a dance in a rehearsal room where male students wear women's heels. But he also performs for Hernández's camera expert solo routines staged in everyday urban spaces outdoors, such as a children's playground. As we follow Cristhian's story, we realize that such appropriated places are as much infused with his subjective meanings as they are conditioned by the objective determinants of others over which he can have no control.

As ever in Hernández these locations are highly aestheticized, consecrated in handsome black-and-white photography (the interviews, on the other hand, are in color). And surely some of these scenes, beyond the expertly rehearsed dance pieces, are staged for the film? For example, when Hernández offers us a sequence of Cristhian meeting with an obliging hipster client, the latter seems younger and more handsome than the typical escort might hope to expect of a customer. I would argue, however, that this gratification of visual pleasure does not denote wish fulfillment or lazy fantasy. Rather, through Hernández's meticulous stylization, the everyday is dignified and eroticized as the marginal is placed lovingly center frame. It is an ethical gesture of respect, of reaching out to an other in his particularity, that is previously unseen in Hernández's more abstracted works.

Testing the limits of an established creative universe, then, the shorts pose a challenge through their technique, thematics, and fragmentary form to the illusory formal coherence and unquestioned cultural prestige of the feature film, the medium through which the director has mainly made his hard-won career.

The Wagers of Art

Julián Hernández's jocular persona on Twitter, not to mention his easy charm and generosity in person, belie the high seriousness of his films. Yet he has complained that in Mexico he is viewed as a maker of films for "queers," while foreigners see him as a director with a distinctive style ("Julián Hernández" 2012). The conundrum is clear. If his finely crafted features are viewed as art film tout court then they can accede to the distinction of high culture, but their distinguishing feature (the investigation of the "homosexual problematic") is rendered invisible.

Hernández has also said (in the interview with me at the end of this book) that he simply puts into his films things that he wishes to see. This avowed scopophilia or desire to look is no doubt shared by the specialist festival audiences, at home and abroad, where he has had such success. Yet new forms of distribution have laid bare the erotic component in such looks with shameless clarity. For example, one website gives precise timings and minute descriptions for each glimpse of actors' body parts in *Yo soy* (you can see the tip of Hugo Catalán's circumcised penis 100 minutes in) ("I Am Happiness on Earth" 2014).

Yet it could be argued that the spectator's libidinal investment in art cinema, whether gay- or straight-themed, is an essential part of the history of the genre. This is especially the case in the United States where foreign auteur films were once less subject to censorship than the commercial genre movies made in Hollywood. We will return to this phenomenon in chapter 4 in the context of the cinephilia of the 1960s, a passionate art film ritual that was inextricable from its fixation on disrobed female bodies.

To his credit Hernández has in his career commented openly on this often-disavowed relationship, situating art cinema within a loving but critical discussion of porn and prostitution. And critics like Subero have praised Hernández's display of mestizo bodies not as a source of visual pleasure but as a political and social intervention in a Mexican culture still riven by caste and where the mediascape remains fatally lacking in ethnic diversity. This social good constituted by a new extension of the ideal body type is not so far from the one found by Podalsky, who identifies newly emergent subjectivities in Hernández's familiar urban landscapes. His treatment of city sites also aestheticizes everyday objects that were hitherto neglected or despised.

One of Pierre Bourdieu's fundamental "rules of art" is precisely this elite aestheticization of the everyday object, its abstraction into a pure form that transcends dull material referentiality (1996, passim). Here I think of the city lights shimmering behind the silhouette of a naked male figure ("Tomás" or "Cristhian") in the final shot of *Bramadero*. Although the ambient sound of the street (the police sirens and organ grinders so typical of Mexico City) still drifts up to Hernández's makeshift location, this remains a consummately romantic image far from the grimy reality of the traffic-choked street below, the Eje Central Lázaro Cárdenas in the Historic Center. In this case, the aesthetic effect comes from distance. Elsewhere it comes from closeness, as when the camera tracks so lovingly and repeatedly over burnished male skin. Truly Hernández's cinema is, in the words of his favorite singer José José, "so near, so far."

Yet Hernández and his team are also well aware (to cite Bourdieu once more) of the cultural field (1996, passim), combining as their project does the aesthetic abstractions of art with the material constrictions of commerce. They have learned how expertly to navigate the unforgiving and fluctuating currents of Mexico's cultural and educational institutions from the CUEC and UNAM via the Cineteca and IMCINE to the festivals of Guadalajara and Morelia. And if, as they claim, those institutions are no longer overtly homophobic (no longer marked by de la Mora's "cinemachismo"), surely this is to a large extent because of Hernández's own track record as the sole internationally acclaimed Mexican gay auteur. Indeed, as we have seen, Hernández is proud of his continuing commitment to shoot in his home country. It is his achievement, then, to have created a queer cinema that is profoundly imbued with everyday existence in Mexico even as it seeks and needs to separate itself from that existence in order to achieve the artistic distinction that justifies its precarious place in the cultural field.

Like Carrillo's gay informants, then, Hernández has employed "scripts" and "strategies" to negotiate a career that is at once amorous and cinematic. It is a unique one in bearing witness, both directly and indirectly, to queer lives in Mexico. And, in a cinematic context, his briefly celebrated boys and men, limned in light and shadow, are homoeroticized heirs to the immortal homosocial stars of the Mexican Golden Age. In Bourdieu's terms, once more and finally (1994, 257), Hernández has made a brave bet (*illusio*) in the game of culture on the possibility of a Mexican queer art cinema, a bet that must have seemed most unlikely to pay off. We shall see in the next chapter how documentarians, including Hernández's producer Roberto Fiesco, have also made an unlikely wager on a yet less plausible topic: the life stories of transgender women in Mexico and beyond.

TRANSGENDER DOCUMENTARY

The issue of *TVyNovelas*, the weekly gossip magazine of Televisa (Mexico's most powerful and conservative media conglomerate) dated January 19, 2015, has on its cover a photo of a smiling bikinied babe with the teasing caption: "Hawk or Dove? Find out!" ("¿Gavilán o paloma? ¡Entérate!"). In an article inside that extends over three (unpaginated) pages, Michelle Trujillo poses, teetering on the highest of heels, in skimpy blue and fuchsia costumes or in a golden gown with a plunging neckline. They are images typical of a magazine that offers glamorous ego ideals for its target female audience of telenovela fans and seductive eye candy for rare male browsers. Yet there is a twist here. The headline now reads: "No secrets! S/He reveals that s/he changed sex" ("¡Sin tapujos! Revela que cambió de sexo").

Surprisingly, perhaps, the article appears wholly positive in tone. Trujillo's modest career credits are given as parts in *Guau* (the long-running LGBT talk show on Telehit, the Mexican music channel affiliated with Televisa) and stage show *Rumba y pasión* (a successful backstage musical based on the glamorous *cabaretera* dancehall movies of the Golden Age of Mexican cinema). Michelle offers useful information about her medical history (hormones at fourteen, surgery at sixteen) and claims she was never bullied at school. She also contends in interview that she was always supported by her parents: her daddy's gift to her on her fourteenth birthday was, she says, the changing of her name and gender on all official documents.

In spite of such apparent openness and modernity, her conservatism and that of her family, a characteristic likely to chime with that of *TVyNovelas*'s traditionalist subscribers, is stressed throughout. Michelle's daddy told her that now she was a girl she would have to follow the same rules as her sister, keeping a strict curfew and bringing any prospective boyfriend home for paternal approval. Not that this seemed to happen very often: Michelle (whose current age is not given) claims to have had only three boyfriends since she was fifteen, educated as she was to be "moral." And when asked if she would like to marry (a legal possibility in Mexico City), she replies that while there are no current candidates in sight, she has the good example of her parents' forty-five-year union to inspire her.

Although overtly respectful of her choice (the anonymous journalist addresses Michelle always in the feminine gender), the piece is based nonetheless on the well-worn strategy of morbid curiosity: stripping away the dowdy feathers of the "hawk" to reveal the dazzling plumage of the "dove" beneath. The truth is thus assumed to be, in traditional philosophical style, that which must first be hidden only finally to be disclosed or unconcealed to the viewer ("No secrets!"). It is a sleight of hand or trial by visibility to which many trans people will be sadly subjected in everyday life and which makes their stories admirably suited to the film or TV media.

In this third chapter, I treat three very recent documentaries on trans women. What is striking is that beyond the broad brushstrokes of "gender as performance," rehearsed by academics especially in the United States for a quarter century now, these feature films address their protagonists' agency in the performing arts (song and dance) and in politics (revolutionary and otherwise) in very specific national and international contexts. Moreover, the geographical reach of the films extends beyond the borders of the Mexican state as far as Cuba or Thailand, even as they remain anchored in the heart of the capital (the grungy impoverished barrio of Garibaldi) or in the everyday provinces (San Miguel de Allende in Guanajuato).

Moreover, as in the case of Michelle's story, so dependent on her parents' support, these transnational narratives break out of the barriers of subjectivity (of a strictly individual history) to embrace an analysis of intersubjectivity (a necessary relation with an other, whether partner, parent, or film director). As we shall see, within the films and in the copious press coverage they provoked (all of which is highly positive), the singular gives way to the universal, the event to the process, and the documentary to fiction. Finally, these are all self-proclaimed stories of love, even as they testify intermittently also to the disruption or affliction (Spanish "quebranto," from "quebrar," "to break") that is cited in the title of one of the features.

General scholarship on trans documentary also attests to conflicts with and within wider queer film studies. Thus, the excellent early collection by Chris Holmlund and Cynthia Fuchs, *Between the Sheets, in the Streets: Queer, Lesbian, Gay Documentary* (1997), does not mention the transgender theme in its title and includes only one, penultimate essay dedicated exclusively to the subject. Yet the volume's introduction (1–14), which does make frequent reference to trans people, raises many relevant points about (trans)gender and (film) genre.

Holmlund and Fuchs begin by proclaiming the "uncoupling of identity (and not just lesbian or gay identity) from any particular aesthetic" (1997, 1), especially now that "firm distinctions among documentary, fiction, and avant-garde films and videos are increasingly untenable" (2). Conversely, however, "generic distinctions do still shape the reception, distribution, and exhibition of documentaries" (2). (Mexican documentarians frequently complain about lack of distribution, in spite of an acknowledged boom in the quality and quantity of their films.) And this perceived yet contested genre is split between what influential scholar Bill Nichols calls "the status of documentary as *evidence from* the world" and the "status of documentary as *discourse about* the world" (in Holmlund and Fuchs 1997, 2). This fracture or rupture between inside (evidence) and outside (discourse) will prove essential to my three Mexican films also.

Under the heading "Queering Documentary," Holmlund and Fuchs list the formal characteristics still held to be typical of the genre and which are rarely fulfilled by the texts studied in their book: "voice-of-God narration, long takes and sync sound, talking heads, archival photographs and footage, interviews, and so on" (Holmlund and Fuchs 1997, 4). But while many of their US queer filmmakers (like the Mexican ones in this chapter) "consciously blur distinctions between documentary and fiction" (4), operating as ethnographers of their own culture (5), still "specific contexts [and] interpretive communities" impact reception (5). Finally, while "no 'straight' lines can be drawn around documentary," queer (lesbian, gay, transgender) examples of the embattled genre "remain narratives grounded in some version of actuality and experience, involving social actors, as opposed to characters" (11). As we shall see, this distinction (actor versus character) will prove to be a vital but slippery one in my Mexican transgender trilogy.

The essay devoted to the topic in Holmlund and Fuchs's anthology is Chris Straayer's "Transgender Mirrors: Queering Sexual Difference" (207–33). Straayer treats two rare independent video documentaries from the early 1990s, one on a "bearded lady" and another on a "transgendered lesbian," both works that "dispute binary sex and the sex-gender matrix" (207). But she stresses here rather the special status of the indie: "I am less concerned with the distinction between representation and reality (the issue of document) than with the competition between different representations to define 'reality' (the issue of independence)" (208). (I will also address the question of to what extent my Mexican films, which typically received government funding and were shown at national festivals, can be called "independent.")

Yet although Straayer cautions against "mistaking representation for reality" still she insists on the testimony her videos give to the "wrenching testimony" of "outsiders" (Holmlund and Fuchs 1997, 208). And if the bearded lady turns herself quite literally into a circus act (juggling and eating fire in the ring [209]), then her director also makes her (unseen) presence felt (narrating on voiceover her own experience of making the film [210]). Straayer reminds us that trans performances, when they go unrecorded by indie documentarians, carry the deadliest of risks. She ends her essay with solemn references to the brutal murders of Brandon Teena (who was born a woman and lived as a man) and Martha Johnson (a drag queen veteran of Stonewall), working-class martyrs to the trans cause (220).

Academic Hispanists' contributions to the field tend to shelter limited analyses behind very broad titles. The prolific pioneer David William Foster's "Documenting Queer, Queer Documentary" treats some much-studied gay-themed features (one of which is fiction) only from Cuba (2010). Gemma Pérez-Sánchez's "Transnational Conversations in Migration, Queer, and Transgender Studies: Multimedia Storyspaces" (also 2010) analyzes a video installation, a documentary short, and a collection of photographic self-portraits, all originating from Spain. Neither article refers to Mexico, much less Mexican trans culture. Nor does a figure such as Paul B. Preciado (formerly Beatriz), a Spanish-born disciple of Derrida whose theoretical writings on the "contrasexual" (2002) do not address film.

However, two semantically rich and emotionally charged visual testimonies to Mexican trans culture and its lived history have recently appeared in print (I bought them in 2015 at Voces en Tinta [Voices in Ink], the queer bookstore in Mexico City's Zona Rosa). Like my three documentary features, these two books both work through the past and stress the pleasures of the present, even when those pleasures are taken, as they so often and so poignantly are, in extremis.

The first text is *Mujercitos!* ("Ladyboys!"), a compilation of press clippings mainly from the 1970s from scandal sheet *Alarma!*, published as part of a research project on this neglected theme by Susana Vargas (2015). The clippings are arranged under four contradictory headings. "Orgy" shows transwomen, generally of low social class and dark complexion who have often been apprehended by the police, as they loll in miniskirts or hot pants or socialize in cocktail dresses (no sexual activity is shown). "Weddings" reveals the *mujercitos*' commitment ceremonies in nightclubs or even prison, with one couple claiming to have been together for twelve years. The accompanying text, fearless of contradiction, finds orgies and weddings each as "disgusting" as the other. "Posing" sees the ladyboys in a variety of staged situations. As Vargas writes in her introduction, Lorena is a modern housewife, smoking as she sweeps the floor or sews; Claudia is a society hostess, with the stiff back of an aristocratic lady; and the glamorous Paulette is reminiscent of Dolores del Río, Mexico's greatest star of the Golden Age and in Hollywood, except that she is shot in a police station not a movie studio (Vargas 2015, unpaginated). Finally, "deception" focuses on trans sex workers who allegedly hoodwinked foolish or drunken heterosexual clients. *Alerta!*'s morbid fascination with the phenomenon it claims to deplore displays the hallowed strategy of revelation we saw earlier in the more modest and apparently progressive pages of *TVyNovelas*: the stripping away of disguise to reveal the supposed truth within.

Picking up on the distinction derived from Nichols that I mentioned earlier, there is, however, a transparent divergence here between "evidence" from the world and "discourse" about the world. By focusing every month for some twenty years on this marginalized group, *Alerta!* gave the *mujercitos* (disadvantaged by class, race, and gender) a national platform with a wide audience (they themselves come from provincial towns all over the Mexican republic, not just the decadent capital). The ladyboys' defiant, often smiling gaze is clearly seductive. Vargas writes unambiguously: "Lorena is beautiful and this photograph shows it." The supposed orgies, weddings, poses, and deceptions are read by Vargas (and Cuauhtémoc Medina, the art historian who provides a preface) as clear examples of resistance to a disciplinary trial of visibility with potentially tragic consequences.

Yet with the passage of time, what has become visible is not only the evidence provided by the book for the spectacle of confident transwomen, rare indeed in the period, but also the hysterical, repetitive tone of the trans- and homophobic commentaries in the accompanying text ("They disgust with their disgustingness"). This disapproving text barely masks the continuing and profound erotic fascination clearly felt by writers and readers of the magazine.

The second book is photographer Susana Casarin's *Realidades y deseos* ("Realities and Desires," 2012), published in a collection by the prestigious magazine *Artes de México* and supported by the Mexico City government and its ministry of culture. The fruit of Casarin's lengthy collaboration with transwomen who are cabaret performers and sex workers, based once more in different locations around the Mexican republic, the photos focus on four handsome brave subjects at work and home. The stress on everyday life is evident in the cover image, a modest washing line where panties, pink and blue, flutter against a gray sky.

Shot in intense, saturated color (so different from *Mujercitos*' smudgily evocative black and white), Casarin's pictures offer once more valuable evidence for transwomen's continued survival in spite of poverty and social exclusion. Indeed, some of the most eloquent images (like that of the washing line) show no human figures. Elsewhere, a pink curtain billows over a modest, empty bed (Casarin 2012, I) or, again, a squalid bathroom, used by the many who spend the night in that flat, is painted a vivid turquoise (VIII). Bearing testimony to an uncharted, neglected subculture, the photos witness solidarity and sociability, as well as marginality: Lenny sits soberly on a couch with the other women in her family (V); a more festive gaggle of young ladyboys sit entwined on another couch, smiling for the camera (IX).

This evidence from a little-known world is qualified by discourse about that world, most clearly in Casarin's commentaries at the end of her book. Thus, the friends on the couch are, we are told, seeking refuge from macho abuse outside the precious haven of their home (Casarin 2012, unpaginated). The brothers of a defiant-looking Agustín in working-class Ecatepec (a crime-wracked city in Mexico State) are said to be "violent" and "aggressive" with the sibling they dismiss as a *joto* (XXXVI). And the final ominous image (XLII) is of barely literate graffiti, blood red on a white wall: "AGUSTIN VETE HO [*sic*] MUERES PUTO" ("Agustín get out or die queer").

Casarin shoots in natural light. Yet she calls her photos "mise-en-scènes," associating them with theatrical or cinematic setups (Montaño Garfias 2009). But this loving, exquisitely crafted technique is not simply a way of dignifying the transwomen whom she portrays with such love, although clearly it is. It is also a sign that (as in the apparently very different case of the *mujercitos*) brute documentary evidence of the world has become a more elaborated discourse on the world that draws strategically on fictional technique. Paradoxically, perhaps mere characters have thus been transformed into genuine social actors. It is a precious conjuring trick attempted also by the three documentaries I treat in the rest of this chapter, all of which are the first features from their respective directors.

Social Commitments: *Morir de pie* ("To Die Standing Up," Jacaranda Correa, 2011)

Irina Layevska, discapacitada y discriminada, se aferra a la vida construyendo su propio personaje. Su lucha comienza cuando abandona la causa social dejando de ser el Ché mexicano, enfrentándose a su lucha personal con Nélida, su mujer.

Handicap[ped] and discriminated [against], Irina Layevska defies adversity and faces life with her true self. Her fight begins when she decides to abandon the social cause, seizing [*sic*, for "ceasing"] to be the Mexican Che, facing a bitter and arduous personal battle alongside her partner, Nélida. (FICG 2011)

The poster for *Morir de pie*, the story of a Communist militant who becomes a transgender activist, suggests the disruption or break that will feature in all the films in this chapter. Only after we see the documentary will we realize that the revolutionary with the dark beard and beret on the left of the image is the same person as the blond with the come hither look on the right; and that the woman in the glasses that is placed between them is the wife of both. The official synopses from the Guadalajara International Film Festival (where *Morir de pie* premiered and won the award for Best Feature Documentary) also exhibit a certain fracture. Thus, while the Spanish original would be better translated "she holds tightly on to life, constructing her own character," the English version claims she "faces life with her own true self." It is a difference, pervasive in writing on transgender documentary, between character and actor, or interior and exterior, which we noted in the introduction to this chapter.

The limited press coverage of the film is likewise ambivalent and contradictory. In an interview in daily *El Informador*, the director is quoted as saying that this is a "universal story for all kinds of audience" (Esparza 2011). Conversely, film blog Homocinefilus stresses the particularity of *Morir de pie*'s political dimension: "People . . . will be able to appreciate different viewpoints on the Left that operates in this country" ("Jacaranda Correa" 2011). In its generally positive review, literary monthly *Letras Libres* (Aguilar 2011) points rather to structural questions in the film's narrative. The plot is based on an element of "surprise" by which the second half of the film cannot be anticipated in the first. And there are confusing "jumps in time" that are not specified or labeled. The critic also has reservations on the extended, perhaps intrusive or voyeuristic scenes in which the protagonist, who has long suffered from degenerative neuropathy, struggles to bathe or clothe herself.

In a lengthy interview with IMCINE, however ("Transita" 2011), the director Jacaranda Correa, a well-known producer and presenter on educational TV Channel 22, stresses her film's links with journalism. It is a connection, she says, that documentary film has not often wished to make. She also cites her difficulty in adopting the right "sensibility" to those whom she describes as her "characters," neither being "soft" nor "offending them." Above all, she says, *Morir de pie* is an "ode to love."

While few spectators who paid to see the film could be unaware of the personal transition to be traced in it, the opening sequences are indeed (as *Letras Libres* noted) willfully confusing. After credits that give notice of support from a range of government bodies (IMCINE, CONACULTA, and FOPROCINE, the agency for "quality production" that gave some 900,000 pesos), the first shot is of hands massaging an obliquely seen face. The unattributed voiceover, which will later prove to come from Irina's faithful wife, muses on life and death as a "personal decision" and on "peak moments" as a time of happiness. We cut to grainy home video of a boy playing with a ball. And again to archive footage of a Che Guevara–like man in a wheelchair. Unidentified talking heads, apparently shot at the present time, make enigmatic pronouncements. For example, one young woman says: "He didn't die. He's just not there." (These informants are identified only in the final credits.)

We then cut back to a lengthy archive sequence of interviews with the unnamed Che lookalike. Here he does give dates as he rehearses his backstory. His militant parents sent him for lonely and perilous medical treatment in Romania (1972) and the Soviet Union (1979). In Cuba he was patronized for his disability, called a "poor little thing." His father abused him, asking: "What is a son like this good for?" In further archive footage, the protagonist proclaims his dedication to the Cuban political project of the "New Man" and to the lifting of the US embargo. As a prominent member of the Mexican Committee of Solidarity with Cuba, which brought oil tankers to the island, he is even shown in a snapshot with Fidel himself.

Further home video shows his wedding to Nélida: the happy couple kiss before a backdrop of Cuban and Mexican flags and the young groom struggles to exchange rings with his already crooked fingers. Meanwhile, the couple's friends praise the wife's capacity for love, while she herself is more ambivalent: slowly she came to realize that the "anger" she felt is based on "mourning for a loss." The first act of the film ends with a return to the archive interview: this time the revolutionary denies the existence of Che's new man ("el hombre nuevo"). Now, he contends, there are only men who are new ("los nuevos hombres").

It is only twenty minutes into *Morir de pie*, then, that we are finally, fully shown Irina, the woman who the bearded militant has become. Blond-haired and whispery-voiced, she whistles for her wife to join her in bed (they sleep apart). And shown spooning together, they movingly declare their mutual love. In spite of the element of surprise here, identified by *Letras Libres*, this revelation of a true or new self is not a coercive one, disclosing or unconcealing a shocking revelation to the viewer. Indeed, in traditional observational style, Correa goes on to show us an ordinary, even mundane, day in the two women's lives. Nélida is now seen working in a Mexico City metro ticket booth, in what must be a considerable letdown from her one-time revolutionary ideals. Irina for her part takes to her wheelchair to investigate a power cut outside their home. Nélida speaks ambiguously in a driving shot of her "apprenticeship" (to her partner's disability or to her transgender status?) and of her "commitment" and "objectives." Correa cuts to a shot of washing fluttering on a line on the roof, the same image of vulnerable, valuable everyday life consecrated by still photographer Susana Casarin.

The losses in that life are not minimized. A lengthy sequence in which Nélida feeds Irina her lunch testifies to the difficulties both had in accepting Irina's increasing disability and dependence. Irina tearfully tells the camera her story as an adult: of love, marriage, and the progress of her illness. And struggling to shower and dress (the bra proves particularly tricky), she is shown for some time naked and vulnerable on screen.

In Irina's subsequent account it was her wife who encouraged her (then him) to "get in touch with his feminine side," a suggestion that they did not imagine would be the start of a process that would become permanent. Yet Irina also claims that a single day was all it took: on August 24, 2001, precisely, she put on a dress, felt "at peace," and introduced herself to her wife as a woman.

Extensive in its account of the medical treatment provoked by increasing disability, *Morir de pie* is strangely reticent on the process of gender confirmation, with no mention of hormones or surgery. Where sex is concerned, Correa and her "character" seem to prefer to stick to the psychological realm. Irina now claims that her (his) Che beard and beret were just part of an alienating image. The role played by Che for the young militant was personal rather than political, an "ideal" father to replace an all too deficient real father. During this sequence Irina is shown having her hair styled at home (*Made in Bangkok* has a similar scene) and checking out her new look in a mirror. The editing suggests that for this historical actor, a revolutionary in more than one struggle, politics is all about costume, whether masculine or feminine. But it is the masculine self-image that is the more wounded: Irina says that when she wanted to be like Che it was in part because the two shared a common experience of disability (for Che it was chronic asthma).

The third act of the film shifts to this new social turn, a new commitment by a new (wo)man, far indeed from that envisioned by the Cuban Revolution. The couple, with Irina of course in a wheelchair, are shown at Mexico City's LGBT Pride on Reforma, a small, fragile couple among the multitude of dancing shirtless boys and towering drag queens. At another, unidentified event, Irina is congratulated for her role in achieving trans rights in the capital. The couple are applauded as Irina's documents are changed to make her a woman and one who is married to another woman. Back at home, movingly once more, we see Irina write by poking at the keyboard with a pen held in her mouth.

Yet in spite of this heartening evidence of a historic social change so bravely and painstakingly achieved, *Morir de pie*'s last section is downbeat. Irina confesses that given the continuing progress of her disease, which now threatens her eyesight, she might prefer to "rest" (we understand she means die) if it were not for her terror of leaving her wife alone. And, in a circular structure that belies Irina's narratives of definitive change and linear progress, we are returned at the end of the film to the opening sequence in which (we now realize) it is one wife who tenderly massages the face of another, cooing enigmatic whispery words as she does so. The individual or subjective story is confirmed as intimately intersubjective, an "ode to love" in which personal transition is wholly dependent on an other.

However, if *Morir de pie* brings, in Nichols's words, fragile, moving *evidence from* the world, it also offers *discourse about* that world. After all, its two protagonists see themselves as political activists above all else, experienced (like the film's director) in informing and inciting audiences. And if Correa uses many of the traditional documentary techniques listed by Holmlund and Fuchs (long takes and sync sound, talking heads, archival photographs and footage, interviews), she rejects one crucial resource: voice-of-God narration. As viewers, we are wholly deprived of the orientation provided by temporal and spatial markers, just as Irina and Nélida were as they embarked on a lengthy and laborious journey for which both proved signally unprepared. It is a complex and conscious technique that will be exquisitely developed in my second trans documentary.

Media Memories: **Quebranto** ("Disrupted," Roberto Fiesco, 2013)

Este documental evoca la memoria y el testimonio de dos personajes: Fernando García, conocido como Pinolito, durante su desempeño como actor infantil en la década de los 70, y doña Lilia Ortega, su madre, también actriz. Fernando se asumió como travesti hace algunos años y ahora se hace llamar Coral Bonelli, ambas viven en la colonia Garibaldi añorando su pasado fílmico mientras Coral asume con valor su identidad genérica. Ambas continúan actuando.

The memory and testimony of two characters: Fernando García, known as Pinolito, who was a child actor in the 1970s and Doña Lilia Ortega, his mother, an actress. Fernando came out as a transvestite, some years ago, and now calls himself Coral Bonelli. They live together in Garibaldi [in central Mexico City] yearning for their past in the movies, while Coral bravely comes to terms with her gender identity. They both still perform. (FICG 2013)

The poster for *Quebranto*, which, like *Morir de pie* premiered at Guadalajara, embodies the rupture or breakage of its title. The face of protagonist Coral Bonelli, formerly child star Pinolito, is split down the middle with the two halves roughly joined together to make a grotesque mask. The official synopses (also from Guadalajara) once more suggest a certain contradiction between the banality of everyday life in a degraded barrio of the capital and the glamor of brief or lasting careers in film and cabaret. Both mother and daughter, the synopsis takes care to warn or specify, continue to act.

Unlike the TV journalist Jacaranda Correa, director Roberto Fiesco is an established figure in the Mexican film scene, having produced not only all of the films of his partner Julián Hernández (which I treated in chapter 2) but also many more by novices or established figures in Mexico, such as the respected master of another era, Gabriel Ripstein. And at the end of the credits Fiesco thanks not only and mainly Hernández (who coscripted and edited *Quebranto*) but also Arturo Castelán (the director of the Mix festival treated in chapter 1, which *Quebranto* opened that year), Ripstein, and Marina Stavenhagen, a former director of IMCINE. The film itself boasts lengthy interview footage with Jorge Fons, like Ripstein a respected senior auteur, who happened to direct the child Pinolito's first film.

Screened around Mexico in the prestigious Ambulante documentary festival (which also provided postproduction funding), Fiesco's film also benefited from circuits of gay cinephilia. *Quebranto* won LGBT prizes at Guadalajara and San Sebastián in Spain (where, in Fiesco's absence, I myself was asked to accept the prize), and it features in a small role a key figure in gay Mexican film culture, Joaquín Rodríguez. A respected journalist who died young soon after *Quebranto*'s release, Rodríguez straddled the divide between art and mainstream that structures much of this book: he played bit parts in both Hernández's *Rabioso sol* and Tovar Velarde's *Cuatro lunas* (treated in the next chapter) and served on the staff of the Morelia festival. He was awarded a posthumous Maguey or queer prize at Guadalajara also.

Although a first feature from its director, *Quebranto* is infused with or embedded in this culture of queer Mexican cinephilia, a characteristic noted in the copious press the film received. In an official interview with IMCINE ("*Quebranto*, la nostalgia" 2013), Fiesco says that his documentary is based on personal contact: first he met Lilia, also known as "Doña Pinoles," and then he discovered that her little dark child star son was now a lanky blond woman. Fiesco praises documentary over fiction, claiming that the former offers "a space for freedom" and "a time for reflection" that the latter cannot provide. Yet he suggests also that his aim is to "play with reality," given that his two "characters" are "living inside a cinematic fiction." Finally, he says, these lifelong bit players have been given starring roles. It is an opportunity he has given them "with infinite love."

In another interview for website Filmeweb.net (Muñoz Tinoco 2013), Fiesco also notes that Mexican society has evolved, showing more respect for "transsexuality." It is a claim he makes elsewhere for the growing acceptance of gay men. Fiesco's love for his subjects, then, is perhaps analogous to Mexico's increasing affection for its varied sexual dissidents. In a further interview with daily *Milenio*, Fiesco extends this empathy to urbanism, saying that his own grandparents lived in a barrio of the Historic Center, which is not so different from the film's Garibaldi (Salgado 2013).

Another article in *Milenio* qualifies this empathy, claiming that *Quebranto* does not "feel sorry" for its subjects (Fiesco 2014). This "honesty," Fiesco says, generates a "connection" with the audience, even if the central pair might seem at first grotesque ("esperpénticos"). Staking a claim, like the director of *Morir a pie*, to universality, Fiesco even argues that their life is "not so different from ours." The journalist, however, stresses the media context of the film, invoking successively and beyond cinema nightlife or cabaret, music from tango to mambo, and Fiesco's own choreographic camera movements, relatively rare in the documentary genre (and similar to those of Julián Hernández's fiction features). Fiesco replies that he sought to treat each scene in a different manner, employing in turn a fixed camera with frontal shots; what he calls "evocations"; artificial staging or mise-en-scènes (the same word used by photographer Susana Casarin); and archive footage, including 35 mm film shot by classic directors like Jorge Fons. Finally, his characters are survivors, living on in a world based on "nostalgia," as is suggested by his own film's so frequent media references.

Testimony to *Quebranto*'s impact at the level of critics if not audiences (in theatrical release it opened on just six screens) is the fact that Leftist daily *La Jornada* published no fewer than three pieces by major critics on this apparently marginal film. Leonardo García Tsao wrote (2013) that it is now "customary" that the best Mexican material seen at festivals is in the documentary genre. And he also calls attention to the specific urban dimension of the film (in contrast, *Morir de pie* typically does not identify the Mexico City *colonia* in which its couple lives). On describing the surroundings of Coral, writes García Tsao, *Quebranto* is also the "portrait of the Mexico City whose buildings are half ruined . . . that lives a contradiction between night and day; a city which can swallow up its inhabitants" (Coral's brother was killed at home in the earthquake of 1985).

Reading the film less as evidence of a world than as commentary on that world, Luis Tovar writes rather in auteurist terms that *Quebranto* embodies two of Fiesco's creative interests or indeed "passions": the desire to "preserve national film history" and an intelligent but sensitive approach to "sexual diversity" (Tovar 2013). The film "synthesizes" these two themes. Indeed, the story of Fernando-Pinolito-Coral, a marginal figure if ever there was one, comes to represent that of Mexican cinema itself: sometimes it stands at a high point, sometimes it hits a low, and sometimes it seems on the point of disappearing altogether, but always it carries on. The true survivor for Tovar, then, is a perilously threatened national cinema.

But the subtlest account of *Quebranto* in *La Jornada* is, unsurprisingly perhaps, by respected gay critic Carlos Bonfil, a longtime observer and supporter, as we saw in chapter 1, of the Mix festival (Bonfil 2013). Bonfil has no hesitation in calling Fiesco's film bluntly "a mix of documentary and fiction." Yet if he insists on the "professional distance" Fiesco takes up from his subjects, he also reads the film in an intimate cinephile context shared by critic, director, and no doubt *La Jornada*'s educated readers. Thus, he compares Coral's fiercely protective but smiling mother first to Anna Magnani in Visconti's *Bellissima* (1951) and second to Giulietta Massina in Fellini's *Notti di Cabiria* (1957). For Bonfil, what distinguishes *Quebranto* from the superficially similar *Morir de pie* is the "incredible vitality" of the aged mother, who is at once vigorous and fragile. In a similar paradox Fiesco is finally proclaimed by Bonfil "the youngest of all the old cinephiles" (Bonfil 2013).

If we turn to the film, this mix of reality and fiction, universal and particular, closeness and distance makes itself clear from the start. Thus, the very first shot stages one of those camera movements mentioned by *Milenio*. A slow tracking shot from right to left reveals the facade of a modest Chinese restaurant (incongruously advertising "tortas" or Mexican sandwiches). There are tacky oriental ornaments in the window, which carefully frames our first view of the documentary's subject inside. She is a stocky, straw-haired, strong-faced woman in a red jacket, flowery skirt, and white lacy blouse. As the camera continues its choreography we hear in voiceover (Coral's lips do not move) her story of the sufferings of the subjects she calls somewhat indiscriminately "gays, transvestites, and *jotos*" in Mexico, an everyday struggle that in her case at least has culminated in happiness (compare the invocation of happiness in extremis that begins *Morir de pie*). Fiesco thus employs a very self-conscious filmic technique on his everyday subject, which serves to dignify or aestheticize the impoverished environment in which that subject survives. But notably he scorns the "surprise" of *Morir de pie*, introducing us at the very start to a protagonist who seems at once proud and resigned.

In the second sequence, we see Coral, now dressed in a tight tank top, laboriously washing laundry by hand on the roof of her building. Fiesco cuts to slow pans over the scruffy skyline behind her or to fluttering underpants on a washing line (a motif now familiar from Casarin and *Morir de pie*). Coral describes her daily routine (but not in sync sound): on a Sunday, it consists simply of watching TV. And she rehearses past tragedies without excessive emphasis or emotion (the death of the brother, her own diabetes diagnosis). Yet we later see her leading a vigorous dance class for younger students. Vitality and stoicism seem to go hand in hand.

Next the camera moves slowly over family snapshots (a mother and tiny son, who is at one point dressed incongruously as a *charro* or cowboy complete with moustache). This montage is set to the melancholy tango that gives the film its title. Returning to the present, the camera roams restlessly over the couple's grimy, cluttered home, stuffed as it is with ornaments and photos, dolls and pill bottles, and swelling walls of VHS tapes. Mother and daughter make up together in twin mirrors and then go off on a walk along a desolate pathway. (Later, in the most touching scene, they will dance as a couple.)

We then return to the historical narrative, this time told by Doña Pinoles, whose garbled words sometimes require subtitles even for Mexican viewers. She tells us how her little boy was obsessed with Raphael, a histrionic Spanish singer of the 1960s. Learning to imitate him from a neighbor's television, the soon-to-be-named Pinolito wins a prize for impersonation, a pathetically paltry 15 pesos. With that he is set for the *carpas* circuit, local variety shows performed in temporary tents.

Pinolito's big break was in auteur Jorge Fons's *Esperanza y caridad* ("Hope and Charity," 1972), and Fiesco expends considerable time on this cinematic primal scene. Encouraged by the director, the child truly fought with another small actor on screen and was left with a scar on the forehead to this day. It is painful to watch Pinolito weep on screen, given the adult Coral's testimony that s/he did not truly understanding what s/he was doing at the shoot.

Yet as mother and daughter stroll together once more through the back lots of Churubusco studios (Mexico's Hollywood), where Pinolito's movies had been made long before, present survival seems to win out over the past successes that were won at such heavy costs. Fiesco even has Coral recreate the desperate walk through the city streets of his fictional mother in the film, cutting back and forward between original and remake. Obligingly, here he offers the former child star the much bigger role of his own mother (played in the original by the well-known star Katy Jurado).

And perhaps psychic trauma can be conjured away by the modest pleasures of everyday life. An endearing sequence (similar to the lunch scene in *Morir de pie*) has Coral making and eating "Mexican eggs" for her doting mother. We then go on the next stage of Pinolito's modest career, as a dancer in a bewildering variety of spangled costumes on the variety circuit in the capital (the snapshots are now in color). Pinolito was forced out of cinema and into theater because the fashion for *fichera* (dance hall prostitute) films put an end to his maternal melodramas. Coral is next shown outside the Blanquita Theater on Lázaro Cárdenas street near Garibaldi where s/he once danced still as a man (the huge theater closed shortly after *Quebranto* was shot). *Quebranto* then offers us a full-length choreographed number from Coral and two of her now mature colleagues. The dance is shot in a single take, showing the performers' professionalism twenty years after they last danced that piece.

Daringly, Fiesco also inserts three full-length musical performances (or mise-en-scènes) into his documentary footage. First, Coral lip syncs to a sorrowful *ranchera* in El Bremen bar in Garibaldi, resplendent in full mariachi gear and fulsomely introduced by Joaquín Fernández. Second, she repeats her child self's success as Raphael, lip syncing once more, but dressed as the Spaniard and this time dancing down the staircase in her scruffy building (this sequence is in black and white).

But, in a third musical sequence, Fiesco also gives Doña Pinoles, professionally made up and coiffed, her own close-up: it is a number sung in her movingly broken voice to the accompaniment of a plaintive accordion and superimposed with handsome images of falling feathers or flowers. By finally offering both his film's longtime supporting players starring roles, Fiesco affirms both his evident love for his disadvantaged subjects, in spite of the professional distance he keeps from them, and his deep affection for and knowledge of Mexican song, dance, and cinema. It is significant that, unlike *Morir de pie*, *Quebranto* has credits for makeup, hair, and production design, a sign of the director's duty of care toward subjects who might otherwise have been presented as grotesque.

Fiesco has noted that Coral did not tell him until late in the shoot that she had also worked as a prostitute. And surprising late scenes show her on the street at night with a group of fellow trans sex workers who discuss the deadly violence that has affected some of their members. Yet *Quebranto*'s final moving sequence simply shows mother and daughter in their shared kitchen confronting, with valor and fortitude, an uncertain future.

The stylistic virtuosity of Fiesco's ending is remarkable. First come night exteriors, shot with a vérité-style handheld, roving camera. Here Coral, wearing for once a brutally short miniskirt, discusses the perils of sex work with the two blonder, flashier girls she calls her "sisters." Shown consistently in group shots and edited with jarring jump cuts, the women wander down a street and pause in front of garishly lit clothing and snack stalls. The latter are known in Mexico as *ambulantes* or "street vendors," just like the transwomen themselves (and, indeed, the documentary festival of the same name where *Quebranto* was shown). In voiceover, Coral tells us of her ambivalence toward her colleagues and details the payments they are all obliged to make to their "protectors."

Fiesco cuts to an ominously empty subway station. Now the camera, no longer handheld, pans and tilts smoothly down to the left where Coral, more conservatively dressed, is seen emerging from the escalator. Shown in tight close-up, she first stares into a shop window and then turns to face the viewer, an ironic or complicit half smile on her lips. The camera pans right to show us what she has been looking at. It comes to rest on a poster advertising the latest Mexican tour by the still massively successful and preternaturally youthful Raphael, the Spanish star Pinolito had first imitated forty years before.

Fiesco now cuts, knowingly, to a monochrome Coral-as-Raphael lip synching the Spaniard's 1968 hymn to the joys of nightlife: "Mi gran noche" ("My Great Night"). Expertly mimicking the original singer's complex choreography, Coral is lit by a markedly artificial spot that swings backward and forward over her face. The song fades out and we hear, for the first time, ambient noise leaching into Coral's building from the street outside. The utopian interlude of the queered song and dance number is over.

Next, the final, moving shot of the film (mother and daughter in their kitchen) is held for a full thirty-five seconds. The camera pulls straight in on the two women, seated and facing forward, as the mother makes a comment about Coral's hair. The mother then looks down, her face briefly cloaked in shadow, and up once more, facing the audience as resolutely as her daughter beside her. Sadly, Doña Lilia was to die shortly after Fiesco's film premiered.

The virtuosity of Fiesco's varied technique in the last five minutes of *Quebranto* corresponds to the many faces of Fernando/Pinolito/Coral. But it also relates to the many attitudes his film so assuredly evokes in its audience: from vicarious fear at street violence, to ironic humor at artistic performance, to profound pathos at the indignities of old age and mortality.

The impoverished modern-day Coral may seem mocked by her brief youthful celebrity. And her preferred Raphael song is, as we have seen, "Mi gran noche." Yet the lyrics to that song suggest that "love is better when all is dark," hinting at a queer subtext to a big mainstream ballad. By bravely stepping out of that darkness Coral has exposed herself to a trial of visibility that, as she knows all too well, can have deadly consequences for trans people. Yet this film by Fiesco, the youngest of all the old cinephiles, is testimony also to the necessary intersubjectivity of a subjective transition. *Quebranto* shows precious evidence of the unconditional love of a mother for a child (both son and daughter) that transcends homo- and transphobia. It is a parental love that my third and final subject will not enjoy.

Songs of Seduction: *Made in Bangkok* (Flavio Florencio, 2015)

Morganna es una soprano transgénero con una determinación implacable en la lucha por romper con los prejuicios familiares y el estigma social, y con la autodeterminación de vivir dignamente reconocida por la sociedad. Con este fin se embarca en una odisea con luces y sombras para tratar de construir la identidad que ha perseguido toda su vida, una identidad hecha en Bangkok.

Morganna is a transgender soprano with a relentless determination to fight against social stigma and family prejudice to the attainment of a universal milestone: the self-assertion of herself as a human being with dignity socially recognized. To this end, he [*sic*] embarks on an odyssey with lights and shadows to try to build an identity persecuted [*sic*, for "pursued"] throughout her life, an identity made in Bangkok.

The poster for *Made in Bangkok* from director Flavio Florencio, an Argentine based in Mexico City, shows its glamorous subject Morganna (raven hair, scarlet blouse) gazing pensively down in front of an exotic backdrop of neon lights (blue, red, and yellow), which signals the modern Asian metropolis.

The image captures the simple premise of the film: a Mexican transgender woman who seeks sex reassignment surgery in Bangkok, which she hopes will be paid for by winning a beauty contest. And *Made in Bangkok*'s narrative and film technique are noticeably simpler than our first two films. Florencio will follow Morganna's "odyssey" sequentially from Mexico to Asia and back to Mexico again with no flashbacks, recreations, or appeal to archive still or moving photography. Indeed, we must resort to press clippings to discover that, as a man, Morganna had trained in classical music at the Guanajuato conservatory. Moreover, this low-budget film has (unlike *Quebranto*) no credits for production design, costume, or makeup, elements that are, ironically enough, extremely important in its pro-filmic world.

Yet, as we shall see, *Made in Bangkok* is perhaps the most complex of our films in its presentation of an identity that is presented as openly constructed (rather than essential or preexisting) and that also takes place in an alien place and in a foreign language (most of the film's dialogue is in English, as is its original title). And while the background of *Morir de pie*'s director is in TV journalism and that of *Quebranto*'s is in film production, *Made in Bangkok*'s is in international film festivals: Florencio had worked at the Zanzibar International Film Festival in Tanzania before becoming artistic director of the Mexican Human Rights Film Festival.

Made in Bangkok is thus a test case for what a Mexican transgender documentary looks like in a global context of production (it was coproduced with Germany) and distribution (it was shown in festivals around the world). Yet still the film depended on familiar Mexican film institutions and subsidies. Post-produced (like *Morir*) with funding from the government scheme FOPROCINE, *Made in Bangkok* won the Press Award at Guadalajara, where it, like my other films, was first shown.

The press coverage tends to place *Made in Bangkok* in the human rights setting suggested by its director's career. Daily *El Universal* begins its review (Huerta 2015a) by noting that the film was especially screened for legal officials in Mexico City in order to "sensitize" them to the issue. When it won best documentary at the Guanajuato Film Festival (considered second rank compared to Guadalajara and Morelia) the same paper wrote that it was "sexual diversity" that had received a prize there, rather than the film itself (Huerta 2015b). Only Carlos Bonfil in *La Jornada* suggested that *Made in Bangkok* now formed part of a Mexican cinematic corpus on the theme (he cites the other two films I treat in this chapter) (Bonfil 2015a), and that it is distinguished not so much by its documentary treatment of a human rights issue but rather by the "free and easy manner" and "challenging vitality" of its subject: a beautiful singer who shows no trace of the homophobia from which she doubtless suffered.

The longest and most sympathetic piece comes in the form of a double interview with director and subject in lifestyle magazine *Noir* (Andreu 2014). The journalist begins by asking the model-handsome Florencio about his background in charity work and how he met Morganna (here spelled "Morgana"). This was an archetypal example of movie star discovery: he found her by chance singing opera in a Mexico City cantina and wondered how she lived during the day "in such a macho city." Moreover, Florencio denies that his intention was to promote general social or humanitarian "causes" with this his first film; rather, he wanted to tell a particular story, with Morganna as star.

The accompanying interview with Morganna herself reveals yet more complicity with its subject. The occasion is that the Mexico City Ministry of Culture has invited her to perform in the *colonia* of La Roma for the benefit of people who are trans and/or living with HIV (the accompanying picture, however, shows her singing to a huge crowd in the central Zócalo, the capital's main square). Morganna playfully claims she has changed "género": both sexual "gender" and musical "genre" (trained as an opera singer, originally a counter tenor, she has now gone more pop). When asked what she brought back in her suitcase from "the trip of her life" to Thailand, she replies, "Myself!" She was "born in Bangkok" and now considers herself a Mexican-Thai singer. Briefly acknowledging the bullying she experienced in earlier life and even the murders of some of her trans friends, she claims that the moral of the film is not just survival but also love (a key theme of the other documentaries in this chapter). Finally, she is asked what kind of men she likes. The answer, given amid laughs and blushes, is "self-confident." And "tall."

What this press coverage shows is that the trans theme has started to shift from a serious social issue to an accepted source of sociability and pleasure. The fact that Morganna, unlike the protagonists of my previous films, is beautiful and passes easily as a woman is clearly an advantage here. Where *Morir de pie* stresses unending political commitment and *Quebranto* a melancholic nostalgia for lost stardom, *Bangkok* suggests rather a positive dimension of self-realization in both life and art. Yet that self-realization is to be found far from home and via a painful surgical process that the film lays out for us in bloody detail. And if Morganna is relentlessly upbeat, albeit sometimes smiling through tears, she is clearly supported by a steely determination to achieve her goals. Here she has much in common with her recent predecessors in Mexican trans documentary, Irina and Coral.

Florencio literally sticks close to his subject throughout the film. In the first sequence, we hear an operatic aria before seeing its source. Morganna is (like Coral in *Quebranto*) framed by a window, in this case one set in the kitchen door behind which she is making morning coffee. We then see her beauty routine as she makes up in the bathroom mirror before the camera follows her walking down the street, her path briefly blocked by a freight train. (The location is San Miguel de Allende, a pretty town favored by US expatriates whose architectural treasures are not shown here.) While having her hair done by a friendly stylist (compare the similar scene in *Morir de pie*), Morganna reveals her unlikely plan of funding her gender confirmation surgery by winning the title of Miss International Queen 2012 halfway around the world.

While these early sequences are strictly observational, the documentary turns participatory or even reflexive as there next comes a first crucial intervention via voiceover from the Argentine-accented director. Florencio exclaims excitedly that this is the first time that Morganna has left Mexico and the first time that he has made a film. The transgender and cinematic projects are thus concurrent and parallel in *Made in Bangkok*. And, unlike in *Quebranto*, the theme of cinema moves from within the diegesis (Coral-Pinolito's career in the entertainment industry) to outside (the making of the documentary that we are currently seeing).

Even more than *Quebranto*, however, *Made in Bangkok* is focused on mise-en-scène. But while *Quebranto* appealed to a shared heritage of national cinema, here, performance is disrupted by cultural difference. Arriving at her Thai hotel (the opening sequences shot in Mexico take up just five minutes of the film), the unfailingly gracious Morganna is unclear whether to drink or wash her hands in a bowl of green liquid she is offered (it is indeed a welcoming drink). She then ventures bravely into the city to hire the hair and makeup specialists she needs for the competition. Much of the film documents the backstage labor required for glamorous runway appearances in gown, swimsuit, and local dress for a very professional competition that will be broadcast live to 160 countries.

It is this (trans)national element that is most distinctive. Morganna soon bonds with her diverse fellow contestants. For example, Miss Russia is deaf, which does not of course stop her communicating with the group. The vivacious Miss Venezuela, with whom Morganna shares a night on the town, is insistent (unlike Morganna) that she wants to hold on to her penis. Another Miss admits to "having a bad time with my transition." Morganna, in a fanciful version of national dress (she wears a huge mantilla), repeatedly rehearses her opening line to the crowd in English: "I come from Mexico, the land of joy and hope." It would be an understatement to say that this is a sentiment rarely expressed in the cinema of her country. The overt cultural nationalism continues in Morganna's choice of a special skill at the contest. While Miss Japan performs as a geisha and Miss Russia as a classical ballerina, Miss Mexico is to sing (beautifully) the classic bolero "Bésame mucho."

Tears ensue when, in spite of her expert performance and handsome appearance, Morganna does not win the contest. Yet here the director will intervene decisively once more. He has sent a tape of the soprano to Bangkok's star surgeon of sexual reassignment, who, luckily enough, is an opera lover. And Morganna's costly treatment will now be offered for free. After lengthy consultation in which we are given detailed descriptions of the surgery (will the vagina be made of penis skin or colon?), Morganna is wheeled into the operating room calling out to her sole friend in Bangkok, the director: "Ciao, Flavio!" He wishes her luck (off screen) in turn. This brief, moving exchange contrasts with a previous sequence where Morganna assures her mother on the phone that "everything is great" in Thailand, but noticeably fails to tell her the momentous news of her forthcoming operation.

Florencio keeps close once more during Morganna's convalescence. On the morning after she murmurs bilingually from bed: "I'm a girl! ¡Soy mujer!" and then admires her new vagina in a mirror (the discreet framing ensures that viewers do not see it). The subsequent pain of, say, learning to walk again, is stressed as much as the final satisfaction of self-realization, much delayed. Yet soon, it seems, Morganna, invited to the event by her kindly surgeon, is singing Carmen's Habanera ("L'amour est un oiseau rebelle") to five hundred specialists in sexual dysphoria on a riverboat in Bangkok. It is a glamorous audience and setting that Coral Bonelli, in grimy Garibaldi, would surely die for.

But like Coral, even Morganna must be reconciled to reality. Back in provincial San Miguel de Allende we suddenly see her changing her appearance in a taxi: a baseball cap, jacket, and sneakers turn her into an unconvincing young man. She tells us (tells the camera) that this disguise is because her father cannot accept her new identity (one wonders if he will ever see the film in which his daughter is so charismatic).

We have followed to its inconclusive end a story that provides moving evidence of the little-known world of transgender beauty contests. And we have seen how Morganna's director serves the intersubjective role played in our earlier films by a wife or mother. But although *Made in Bangkok*'s star and filmmaker seem reluctant to present their work in the context of a social cause, much less as a denunciation of transphobia, this final, unexpectedly downbeat sequence suggests that Morganna and her transnational sisters, who have come so far already, still have a long way to travel toward universal acceptance.

Lives in the Mirror

So new is the transgender theme that these films have no stable vocabulary in Spanish to address it. Irina says she became a "mujer" ("woman") when she first put on her wife's dress; Coral calls herself "gay" or "joto" and is identified by the film's synopsis as a "travesti"; in Morganna's discussion with fellow Misses at the beauty pageant "transgénero" seems to be used for pre-op and "transexual" for post.

In similar style the first two films are noticeably coy about their subjects' transition. Only *Made in Bangkok*, with its lengthy clinical scenes, goes into detail about Morganna's use of hormones, earlier orchiectomy (testicle removal), and diverse options for final surgery. Perhaps there is some displacement at work here to, with each film preferring to speak about something other than its main topic: *Morir de pie* dwells on politics and disability, *Quebranto* on cinema and theater, and *Made in Bangkok* on culture shock (what is that green liquid presented to Morganna by the ever-smiling Thais?).

The films might also be accused of playing down the deadly violence faced by some trans women in Mexico, a toxic inheritance from the *mujercitos* of old that echoes down to Susana Casarin's contemporary regional sex workers. We learn more context only subsequently. *Morir* the film does not tell us that Irina's transition was received with hostility by her neighbors. Locals' rejection of Coral is voiced in *Quebranto* by her loving, heartbroken mother, not by the protagonist herself, who claims she is "happy." Morganna recounts only late in *Made in Bangkok* that after being bullied at school she would take her time walking home so that her parents would not see her tears.

But these documentaries are caught in a double bind. If they stress the universal they promote audience identification with their subjects (suggesting, as Fiesco said, that "their lives are not so different from ours"). But they may then lose particularity, weakening the invaluable evidence they present to a majority audience from the little-known world of a marginalized minority. Conversely, if the films stress the particular, they problematize identification (surely the extraordinary lives of Irina, Coral, and Morganna are not like ours?) and limit the possibility of employing a strategic universality. The films would thus weaken their vital potential role as discourse on a world that serves to promote our interconnection with others in a common humanity. Yet, in spite of their disruption of stable distinctions between singular and universal, event and process, and (most especially in the assured *Quebranto*) documentary and fiction, they still remain (as Holmlund and Fuchs wrote of their US films) narratives grounded in some version of actuality and experience, involving compelling social actors or agents, as opposed to mere characters or fictional figures.

Likewise, in spite of so much blurring, documentary remains distinct as a genre, with special conditions still shaping its production, distribution, and exhibition. *Quebranto* benefited from admission to Ambulante, a now well established traveling festival in the field. All three films premiered at Mexico's largest festival, Guadalajara, and received funding from government agencies aiming to promote "quality" cinema in the country. It would seem, then, that there is a parallel between the trans theme (readily accepted by the Mexican press) and the documentary genre (eagerly promoted by those same journalists). We have seen that the low-budget documentary on trans women, a previously unknown phenomenon, was rapidly incorporated into Mexican film culture, in spite of being doubly disabled by cinematic genre and (trans)gender theme. The ease of this incorporation raises questions about the relationship between independent (but publicly subsidized) cinema and mainstream movies that I will treat in the next chapter.

It is striking that these three films, which were released so close together, are very different in technique. *Morir a pie*, which never identifies its multiple times, places, and witnesses, is the most disorientating, especially in its appeal to surprise (the bearded revolutionary becomes the blond trans activist). *Quebranto* is the most self-conscious, with its painstaking recreations and elaborate musical numbers juxtaposed with a moving micronarrative of vulnerable everyday life. *Made in Bangkok* is the most minimalist, with the director (in spite of his periodic interventions) just keeping his camera close to the charming Morganna as she goes on the trip of her life. We thus see, as Holmlund and Fuchs wrote once more, that there is an uncoupling of identity (and not just lesbian, gay, or trans identity) from any particular aesthetic in film.

This fluidity is inadvertently reconfirmed by the title of the *TVyNovelas* article on the young transgender actress with which I began this chapter. "Gavilán o paloma" ("Hawk or Dove") is actually the name of a ballad from the 1970s by José José (the romantic singer whose music is featured in Julián Hernández's films). But rather than referring to a fixed, true identity that must be revealed beneath false feathers, the lyrics of the song suggest the instability of the self in the overwhelming experience of love: the singer first believed he was a proud hawk, seducing the beautiful lady to whom he sings, but has since been reduced to her humble servant, a trained dove eating out of her hand.

What the three documentaries have in common, however, is that an apparently subjective story is presented as intimately intersubjective, an "ode to love" for a wife, mother, or filmmaker. This is confirmed, unexpectedly perhaps, by a motif present in all three: the mirror. Irina checks out her appearance after the hair styling, Coral makes up alongside her mother, and Morganna follows her beauty routine in the bathroom. All are preparing for the perilous trial of visibility they will face outside the safe space of their homes. What these films propose, however, is that this self-imaging is not or not only the recognition or reaffirmation of a new subject (a new woman of whom Che could only dream) but also the recreation or construction of a brave life that is made of necessity in collaboration with others.

MAINSTREAM MOVIES

A medium shot of two women in a movie theater, one older with a shorter, dark hairdo, the other younger with a flowing blond mane. The former is concentrating earnestly on the unseen screen (we hear only lugubrious dialogue in German), while the latter initiates attempts to chat, share popcorn, and, finally, kiss her partner. After at first rebuffing these overtures, the older woman gives in and the pair share a passionate embrace. The camera cuts back to extreme long shot. And we see, in an unexpected sight gag, that the couple are almost alone in a large auditorium, which is empty save for scattered solitary foreign film fans.

This sequence comes some twenty minutes into the running time of Raúl Fuentes's lesbian romance *Todo el mundo tiene a alguien menos yo* (Castro Ricalde [2015] examines some other rare films on this theme). And what is striking is, first, that the sequence juxtaposes love of film with love between women, and, second, that the open display of the love between women is fully integrated into the love of film. Thus, the movie theater hosting an obscure foreign movie seems as if it were made to measure for the amorous female couple.

In this fourth chapter, taking my cue from this funny, sexy sequence, I examine three fiction features that I call "mainstream." By this I mean that they are relatively accessible in their aesthetic (although *Todo el mundo*, with its black-and-white cinematography and asymmetrical framing, hews closest to the auteur mode) and also apparently unchallenging in their politics (although *Todo el mundo*, once more, somewhat disconcertingly matches a mature woman with a schoolgirl). Where, as we saw in chapter 2, an art house director like Julián Hernández confronts the spectator with artistically and thematically demanding and original material, these three films follow the templates of familiar genres (romantic comedy and thriller) and promote an unapologetically assimilationist politics (in which queer love is, in the tag line of one of my movies, simply love). This chapter asks, what does it mean for an LGBT cinema to be no longer marginal or subcultural, to be always already integrated (in the audience's fantasy or the producers' intention at least) into the society that has so long and so violently excluded it?

This new phenomenon is surely responsive to social change in Mexico since 2000, when (as I wrote in the introduction to this book) the once hegemonic PRI party was expelled from office after more than seventy years in government. A nationalist, revolutionary ideology, at once political, economic, and cultural, which was often explicitly patriarchal, was finally called into question. And a new more diffuse project (often called neoliberal) unsteadily took its place. Judicial and social changes for LGBT Mexicans came in the wake of what was known simply at the time as "the change" ("el cambio") at the start of the millennium. As also mentioned in the introduction, marriage equality was won in Mexico City as early as 2010 (it remains unavailable in the great majority of the other Mexican states). And the Supreme Court ruled that once ubiquitous homophobic slurs were not protected free speech in 2013. Homo- or transphobia have clearly not been abolished in Mexico (indeed, Mexicans frequently lament their supposed prevalence in the country). But my mainstream films assume a social context in which queer visibility, including the open display of affection (same-sex kisses in a movie theater) is relatively normalized.

My corpus includes lesbian, gay male, and trans fiction features that all gained limited theatrical release after premieres at prestigious local festivals. Their titles are: *Todo el mundo tiene a alguien menos yo* (Raúl Fuentes, 2012), *Cuatro lunas* (Sergio Tovar Velarde, 2014), and *Carmín tropical* (Rigoberto Perezcano, 2014). How, then, to read these new films when they no longer present themselves as resistant to a dominant cinematic and political regime, as was previously the case with much queer cinema in Mexico and beyond? I propose to examine them within three distinct but related prisms.

The first is suggested by my opening scene in the art house: cinephilia. In his 2003 book on the subject, the prolific cultural commentator Antoine de Baecque traces both the "invention of a look" (9) (that is, a new way of seeing cinema as text) and the invention of a culture (18) (a new way of experiencing it as ritual) in Paris after World War II. If cinephilia posited an active mode of viewing, that mode was based on transmedia avant la lettre: the printed text of critics (who soon became filmmakers themselves) had the same vital status as the films they championed or excoriated (De Baecque prefers the epithets "heterogeneous" and "intercontextual" to describe his new social and cultural practice [15]). Cinephilia, in spite of its later reputation for elitism, was also initially transgeneric, prizing and promoting popular US genre cinema that was hitherto despised by an anti-American cultural establishment that supported "quality" French films (18–19). Indeed, de Baecque reminds us that the early countercultural critics (such as André Bazin or François Truffaut) had no institutional position within the French intellectual or academic hierarchy (23).

Finally, although cinephiles paraded their erudition (assembling exhaustive lists of, say, American B pictures reclaimed as film noirs), they were motivated not by reason but by passion, by a love or even lust for cinema. De Baecque cites as decisive their early desire for a young Bergman beauty, Harriet Andersson, and their identification with Hitchcock's mature voyeur in *Rear Window* (25–31). However, this celluloid love is for a lost object. Where once critics sought simply to tell the history of cinema, now de Baecque explores the history of that history (11) in a project marked by melancholic reflexivity (12). As we shall see, cinephilia will also have a distinct inflection for queer viewers, most especially in modern Mexico. Certainly, the Mix Festival and the Mecos porn productions, analyzed in my first chapter, will have helped to nurture a (homo-)eroticized practice of audiovisual consumption located (like Parisian cinephilia) in a distinctly and distinctively urban time and place.

My second prism is perhaps the current equivalent of the cinephilia of the postwar era: fan studies. This discipline, precariously established in Anglophone scholarship, is almost unknown in the Spanish-speaking world. There is no equivalent of Henry Jenkins's foundational monographs on convergence (Internet creativity by movie and TV fans) and "spreadable" media (audiovisual texts produced in expectation of and even in collaboration with obsessive consumers) (Jenkins 2006 and 2013, respectively). In fan studies panels at the 2015 meeting of the Society of Cinema and Media Studies (SCMS), I noted similarities with debates over cinephilia. Thus, fan production (like that of cinephiles) is an activity based on a new way of looking, and one that carries with it its own history. Transmedial in focus, it extends or "spreads" TV and cinema texts into the new creative genres of fanfic and fanart, vindicating once despised or neglected genres, often with a queer twist (a recent favorite of fan studies scholars is the basic cable drama *Supernatural* [CW, 2005–], the springboard for its fans' incestuous "slash" fiction between the series' brother protagonists).

Finally, fan studies prizes erudition but is founded on passion. Young academics at SCMS voiced concerns about whether they could fulfill the twin functions of scholar and enthusiast, without betraying either of the two roles that they deemed to be essential to their research. Indeed, some feared that their careers might be harmed by allegiance to a new discipline less authorized than the related specialisms of audience or reception studies. The question of distinction raised originally by cinephilia (the assertion that B movies could be as culturally canonic as classic French novels) is played out once more in a new field, in a new quest for legitimation. Novice directors of Mexican mainstream film will also appeal (like US TV fans) to social media, even as they seek nonetheless to authorize their practice through the more traditional mechanisms of participation in festivals and graduation from film schools.

My final prism is the most theoretical of the three, but perhaps the most relevant to Mexico, that of posthegemony. While the European nation-state consolidated its power through consensus (the Gramscian definition of a hegemony that is not experienced as violent), in Latin America the state was often weak, based on brutally broken social contracts. For Jon Beasley-Murray the epitome of this was the *requerimiento*, the text ritually read by Spanish conquistadors before they appropriated by force the land and bodies of uncomprehending indigenous people (2011, xix). Like cinephilia and fandom, then, posthegemony is an active practice, indeed in origin a performance that effects a change in the world it claims simply to represent. Evidence for posthegemony is to be found in the transmedia materials Beasley-Murray treats from historical chronicles of discovery and conquest (even Columbus broke his contract with his crew) to novels and television. And like cinephilia and fandom once more, posthegemony is based not on reason but passion: for Beasley-Murray habit and affect (ix) are the fragile foundations of Latin American societies and the means through which social change now takes place. And posthegemony appeals not to the people or the working class as historical agents but to the multitude as a provisional collective grouping. This, I would argue, is reminiscent of the loose but extensive associations of cinephiles, film fans, and (hitherto invisible) LGBT citizens that have recently emerged in Mexico.

Resistance is thus highly problematic when power has historically been at once widely diffused and, unlike in contemporary Europe, overtly violent. Hence, Beasley-Murray chooses not to focus on much-studied rebel movements such as the Mexican Zapatistas of the 1990s, once held to be promising sites of counterhegemony. Yet even in a context where the subaltern can barely speak (the messy politics of the post-PRI era in Mexico might be emblematic of this) "something always escapes" (xxi), eluding the posthegemonic fix. In this chapter I will argue, finally, that that something is the gift, an ambivalent motif present in the plots of all three of my films and one that transcends the logic of economic exchange implicit in any cultural product, however marginal it might appear.

Moreover, with the death of both hegemony and counterhegemony (a nationalist revolutionary patriarchal project and an organized resistance to that project), homosexuality might reappear as, ironically enough, a new focus for consensus in modern Mexican cultural production. It is striking that within the diegesis of these films there is little space for either homophobia or activism. And outside their story space, it is notable also that, just as authoritative critics favored the Mix Festival from its very start, so they lent their support to these mainstream LGBT films, which provoked little or no press controversy.

Lesbian Romance: *Todo el mundo tiene a alguien menos yo* ("Everybody's Got Somebody . . . Not Me," Raúl Fuentes, 2012)

Alejandra está harta de la cotidianeidad y de las relaciones pasadas que no han funcionado en su vida hasta que conoce a María, adolescente con la que empieza una aventura. Al principio todo marcha sobre ruedas, sin embargo, la personalidad de Alejandra y sus exigencias afectivas, resultan cada vez más demandantes al punto de que es imposible estar cerca de ella. En este momento, ambas se plantean si deben continuar con esta relación en la que se padece un opresivo equilibrio o mejor cada cual sigue con su existencia ordinaria.

Alejandra is a successful executive who is tired of her daily routine and her emotional relationships that haven't worked out. One fine day, Maria, a teenager full of energy, crosses her path making her change her expectations. Together, they become involved in an affair based on Maria's admiration for Alejandra's intelligence and experience. At first everything goes smoothly. However, Alejandra's personality and her romantic requirements become increasingly demanding, to the point where it is impossible to be around her. (IMDb 2012)

Todo el mundo, the first and to date only feature by CUEC film school graduate Raúl Fuentes, begins with an enigmatic driving sequence. Because of the framing, we see only an empty seat beside the face of a driver who remains invisible. Soon, however, we cut to a scene of kissing and bare breast caressing in the parked car, all set to an indie pop song with jangling guitars. The lyrics speak of distance and new proximity ("A hundred thousand miles come closer every day").

It is interesting to compare this stylish opening with that of another first queer feature shot in black and white by another CUEC alumnus, Julián Hernández's *Mil nubes de paz* ("A Thousand Clouds of Peace," 2003). Indeed, the directors share personnel. Andrea Portal, as the older woman Alejandra, would soon take a major part in Hernández's fourth feature, *Yo soy la felicidad de este mundo* ("I Am Happiness on Earth," 2014). The high-contrast cinematography of Jero Rod-García (which here won him an Ariel or Mexican Oscar) would grace Hernández's documentary short *Muchacho en la barra* ("Young Man at the Bar," 2015). And the experienced Roberto Fiesco, Hernández's partner, served as an associate producer to the novice Fuentes. Yet while *Mil nubes*'s first scene also boasts a pleasurable soundtrack (a romantic ballad once more from José José, the one-time "prince" of Mexican song), Hernández's number serves as a bitterly ironic commentary on his image track. *Mil nubes*'s protagonist is a rent boy who, having paused to spit out the cum from an anonymous john, sits morosely and lengthily beside him as he is driven home in silence.

Fuentes's female lovers, however, are mutually attracted and animated, rejecting art movie anomie and troubled only by ordinary arguments. And in the second sequence we are offered a glimpse into that everyday affair, before even the social and narrative context of the lovers is established. Languidly postcoital (with bare breasts prominent once more), the women bicker mildly over the smoking of María (Naian González Norvind, billed as Naian Daeva). Later they read aloud in English a text by Samuel Beckett, and María shows Alejandra her paintings for the first time. In spite of the transparent visual and auditory pleasure of the scene (the location will later be specified as Alejandra's elegant modern house in the comfortable Del Valle *colonia*), the cinematography and editing draw still on once disruptive art house traditions. Thus, Fuentes offers the long take that allows us to observe uninterrupted the interaction between the lovers, and he even uses the transgressive technique identified by de Baecque with the origins of cinephilia when Parisian critics lusted for Bergman's teen lover: the look directly into the camera that signals the breaking of the fourth wall. Fuentes's camera observes Daeva, his young half-Scandinavian star (the daughter of a well-known actress of Norwegian origin) with as frank a desire as Bergman looked at Andersson some sixty years before.

The twist here is that the look of love (on and perhaps off screen) is not heterosexual but lesbian and there is no *Rear Window*–style male voyeur within the fiction. Yet the first half of Fuentes's film stages a kind of social and cultural pedagogy analogous to that of the original Parisian version of cinephilia, laying bare the invention of a look and the history of a culture. Young María would seem to be a willing pupil here. Fuentes cuts from the bedroom to the schoolyard where the teenager is dressed in her modest (provocative?) uniform, a first slight sight gag for the spectator who previously had no clue as to her real age and status. And in this posthomophobic world, María's young friends are unfazed by her older lover's attentions (even her parents will, somewhat ambiguously, thank Alejandra for a "friendship" with their daughter that involves frequent overnight stays). I have already mentioned the movie theater kiss as a scene of cinephilic ritual or education, suffused by eroticism. Alejandra will next teach her young partner how to put on makeup (soon smeared once more by a kiss), telling her that lipstick should be shared only "among intelligent people." And they will then dance expertly together at a jazz club that, surely, the teen would be unlikely to have attended before meeting her more experienced partner.

The choice of music, like the juvenile smoking, the straight-to-camera looks, even the retro wardrobe (at one point María sports a jaunty trilby), is knowingly reminiscent of the French New Wave films that arose in tandem with Parisian cinephile criticism. And *Todo el mundo* would seem to confirm the prestige of European references that are high culture in a Mexican context (we saw that cinephilia sought in France to lionize popular genres). Thus, we learn, albeit only some forty minutes into the film, that Alejandra works in a publishing house (it is striking that, eager to situate her professionally, the English-language synopsis, unlike the Spanish, specifies at the start that she is an "executive" and spells out the psychological motives of the couple's attraction). Allied with an archaic print culture, the older lover will even pick up another self-assured and attractive woman in a bookstore, the Porrúa branch in picturesque Chapultepec park. Conversely, when Alejandra attempts to explain abstract art to María in the Tamayo Museum that is located in the same central park, the young girl will take a phone call from a friend, the sign perhaps of her unthinking impatience with the rituals and ceremonies of high culture. The older woman's preference for British indie pop and UK TV is also presented, implicitly, as a culturally distinctive choice, contrasted in its melancholy and solitude with the blaring techno at María's drunken, druggy teen parties.

Yet pedagogy cuts two ways. Alejandra is so technologically clueless that she confuses María's iPhone with an iPod. María, meanwhile, justifies her drug use by citing William Burroughs (the English of both lovers is impeccable). And in the second half of the film, the women's roles are inverted. Slightly disorientating the spectator with a temporal sleight of hand, Fuentes cuts back, halfway through the running time, to the moment the two women meet. The pickup here between the precocious girl, smoking in her school uniform once more, and the reserved older woman, takes place in a somewhat incongruous bowling alley where, characteristically, Alejandra has gone to play on her own. Yet teenage María is no innocent; she is clearly an active participant in a seduction process that pivots on conflicting readings of Plato (María attacks his banishment of the poets, Alejandra cites his praise of amorous "rapture").

Moreover, Alejandra, stricken later by passion and jealousy of Proustian proportions, proves progressively incapable of the social proprieties required by the rituals of cultural distinction she is trying to teach María. Thus, after instructing María on what to order in a fancy restaurant, Alejandra showily spoils the couple's Sunday lunch when she insults a young friend of María's who dares approach the two women's table. And, unable to start her car, Alejandra suffers the indignity of getting a ride from María's tolerant (too tolerant?) parents. It is telling that when the couple finally breaks up, we hear only Alejandra's side of a phone conversation as she lectures María once more on how, logically, they have to be together. In another long take, the camera drifts up through Alejandra's minimalist open-plan home to linger pointedly on the now-empty double bed. Soon Alejandra will be making embarrassingly drunken 4:00 a.m. visits to María's family home. It would seem that the callow teen is more adult than the experienced professional who is at least twice her age.

The definitive break between the two women is signaled at the end of the film by an aesthetic and cultural choice once more. María rejects a book called *Love Stories* offered her at the school entrance by Alejandra, just as, much earlier, Alejandra had rejected María's gift of a volume of Pessoa, a poet who is not to her too-sophisticated taste. And María has cut her blond tumbling hair into a more sober Jean Seberg–type bob, ironically closer to that of the older lover she has rejected. She explains (lies?) about her rejection of the gift: "I don't like to read any more." Yet soon the sullen Alejandra is being picked up once more by a dangerously young girl among the music shelves of a branch of Mixup, the well-known video and record store. It seems that the lesbian pedagogy, still associated, like cinephilia, with the archaic media of the French New Wave (print, vinyl, celluloid) will not stop anytime soon, even after the swift-paced ninety minutes of the director's debut have come to an end.

In a short video made to launch the film, Raúl Fuentes rightly stresses the rarity of the lesbian theme in Mexican film (although, rather than use that word, he prefers to talk of "women in love with women"). Yet the video is sponsored by IMCINE, the national film institute. And the credits to his film cite Conaculta (the ministry of culture), the UNAM (the national university), and the CUEC (the university's prestigious film school, which funded the film through its annual competition for debut features). Fuentes himself thanks Jorge Ayala Blanco, perhaps Mexico's most prestigious film critic, for his help. Within the relatively small world of officially sponsored Mexican culture, necessarily limited by budget and audience when compared to commercial film and TV, much less the Hollywood blockbusters that dominate the multiplexes,

Todo el mundo is thus a mainstream movie. Although it cites the once countercultural techniques of the French New Wave, those techniques have now become reflexive, not so much a way to represent the real as a way to invoke a cinematic representation of the real. In the same way, we saw that cinephilia (in the critic de Baecque and, I would argue, in the film school graduate Fuentes) has become a reflexive project, investigating the invention of a look that is not now new and the history of a culture that, once transgressive, has become institutional.

Further evidence for this queer institutionalization lies in the official promotional video for the 2013 edition of the Mix Festival, whose main home is at the Cineteca Nacional. This was shot at Baños SoDoMe, the first bathhouse founded explicitly for Mexico City's gay men (traditional steam baths had been adopted only latterly by them). Here Andrea Portal (Alejandra in *Todo el mundo*) will feature once more among a group of pansexual frolickers as a sullenly sexy older woman flirting with a more exuberant young girl.

In a similar way, the lesbian theme of Fuentes's film is perhaps no longer countercultural. While it is true that the topic remains rare even in Mexican art cinema, as we shall see in chapter 5, it is not so uncommon in television, which reaches an immeasurably wider audience. Yet *Todo el mundo* does restage that tension between high and low culture first rehearsed by the Parisian cinephiles who lionized a then-despised Hitchcock. After all, it is by no means clear that Alejandra's elite choices are the ones to follow, as she is clearly an unsympathetic character and the architect of her own (intermittent) solitude. If artistic distinction is so misanthropic (María asks her lover: "Why so much hate?"), then the messy, friendly animation of a teenage party seems rather more attractive than Alejandra's customary solitary trips to the Cinemex multiplex ("Just one person").

Yet, like cinephilia once more, *Todo el mundo* shows that passion can coexist with erudition, with the film framing its sexy love scenes with studious New Wave reminiscences. Fuentes is clearly a film fan. And his project, a calling card to the world of Mexican cinema, fuses, like the work of the fan studies scholars I cited in the introduction to this chapter, a transparent enthusiasm for its object with a professional care for formal conventions. Moreover, Fuentes lays himself open, as a male director of a lesbian love story (along with his straight male viewers), to suspicions of a more personal libidinal investment in the sex scenes that he stages with such artistry and sensuality.

Finally comes the question of posthegemony. As mentioned earlier, lesbianism (relatively rare in Mexican film) is revealed here, in a middle-class metropolitan setting at least, as uncontroversial, taken for granted. Young María is not bullied by her classmates (indeed the problem is that she proves too popular with her fellow youngsters for her older lover).

Nor is she shown to come out traumatically to parents who are little seen but apparently supportive of her love life. Alejandra for her part experiences no tension in a workplace that is barely shown. There is thus no repression, no hatred (other than Alejandra's distaste for the teen friends of her lover of whom she is obsessively jealous). But nor is there a space for revolt or resistance. After all, what would María rebel against when she has so little cause for complaint, other than an overattentive, bad-tempered girlfriend?

Yet we might also argue that *Todo el mundo* is, like posthegemony, active or performative. By presenting a perhaps utopian world where lesbianism is indifferent, Fuentes calls that world into being, cloaking it in monochrome cool for his knowing audience (infrequent enigmatic intertitles serve as cinephile *requerimiento* or ritual address). And the director's strategies here are habit and affect. The Mexico City in which the women move and love is at once familiar, everyday (those recognizable movie theaters and bookstores) and exceptional, aestheticized (those stylish interiors and wardrobes). Habit naturalizes love between women, however different their ages, while affect engages the spectator in their passionate and, finally, pathetic story.

As Beasley-Murray suggested, perhaps in Latin America it is through posthegemonic habit and affect, rather than through Enlightenment social contract and reason, that political change takes place. It follows, however, that (as signaled by the first-person pronoun in the film's title) social change presupposes a loose collective of individual cinephile viewers (or, indeed, of lesbian lovers). This spontaneous grouping is an atomized multitude unable or unwilling to organize itself as a historical agent, as was intended by the collective people of past national projects.

Ages of Gay Man: *Cuatro lunas* ("Four Moons," Sergio Tovar Velarde, 2014)

Cuatro historias de amor y autoaceptación: un chico de once años lucha por mantener en secreto la atracción que siente por su primo. Dos amigos de la infancia se reencuentran y comienzan una relación que se complica por el miedo de uno de ellos de ser descubierto. La relación de años de una pareja gay se ve amenazada tras la llegada de un tercero. Un anciano, casado y con hijas, se obsesiona con un joven e intenta reunir el dinero para costear la experiencia. (Filmes&Mas 2015)

Four stories about love and self-acceptance: An eleven-year-old boy struggles to keep secret the attraction he feels toward his male cousin. Two former childhood friends reunite and start a relationship that gets complicated due to one of them's [sic] fear of getting caught. A gay long-lasting relationship is in jeopardy when a third man comes along. An old family man is obsessed with a young male prostitute and tries to raise the money to afford the experience. (IMDb 2014)

While *Todo el mundo* focuses obsessively on a relatively isolated single lesbian relationship, *Cuatro lunas* cuts backward and forward between four gay couples, placing each quite precisely in its collective context. Thus Sergio Tovar Velarde's first feature begins with the following brief sequences: thirty-something Hugo and Andrés argue as they walk back from a movie theater (that cinephile motif once more); tween boys Mauricio and Oliver discuss video games in the school yard (the latter casually uses the homophobic slur "joto" here); retired professor and one-time poet Joaquín looks lustfully (fearfully) at hunky sex worker Gilberto in the Turkish baths; and students Fito and Leo, once best friends back in the provinces, meet up again in a Mexico City university. Tracing the four ages of gay Mexican man, the script relates each to an astronomical metaphor: the moon, which is, according to the life stage of the protagonists, new, waxing, full, and waning. More humorously the film's poster shows four pairs of underpants, each age appropriate to *Cuatro lunas*'s varied gay lovers.

Much more explicitly than *Todo el mundo*, *Cuatro lunas* presents itself as a pedagogy in gay existence, but one that once more is inseparable from good judgment and cultural distinction. Thus, the tensions in the longtime relationship between Hugo (a sophisticated Spaniard) and Andrés (a more modest Mexican) are played out in terms of taste: they next argue over the choice of a fine bottle of wine. And their main source of contention will be Andrés's collection of fridge magnets, a tacky anomaly in the stylish minimalist home the couple share but is apparently owned by Hugo. Young Mauricio longs for the latest version of a video game, all the better to seduce his cousin. Aged Joaquín, obsessed with the sex worker, is scorned in the baths but honored by an homage in a minor university to his little remembered, but surely exquisite, poetry. Fito's widowed mother is addicted to the popular telenovelas that, initially at least, provide an excuse for not knowing about her son's gayness. As they sit warily watching the set she asks him not to tell her "that thing [she] does not like."

Cultural distinction is thus always already a question of conflict or consensus around a fluid gay sensibility. The magnets form part of Hugo's distaste for his partner's supposed effeminacy (though it is Hugo who prefers to bottom in sex). Taking the female role in the video game is dismissed by one child as "gay." The students' first date takes place in a restaurant in emblematic Oscar Wilde Street, located in the old money *colonia* of Polanco, although Fito lives in straitened circumstances. It is only as they discuss favorite songs and books in the tony eatery (according to its website, Brassi boasts "Mediterranean food") that the inexperienced young men start to discover a mutual love for which they have, as yet, no name. Beyond the rigors of exchange value exemplified by sex work, gay pedagogy is identified, as in *Todo el mundo*, with the gift. Leo offers Fito a mix CD made to measure for his new lover's musical taste. And the gift box also contains a card with a message of perfectly and constantly reciprocated love: "Whenever you read this, you can be sure I will be thinking of you at that exact moment."

Properly cinematic in spite of its wordy script, the film focuses on the invention or discovery of a gay look. This look may be explicitly erotic (the married professor eyes the teasing male prostitute as the latter lets his towel gape open) or more broadly sensual (the young boy stares at and finally drinks from the glass of milk his cousin's lips had just touched). But *Cuatro lunas* also charts the invention of a tradition of gay love that is (implausibly, perhaps) initially unknown at least to the innocent students (the tween boy, on the other hand, reveals to a priest in confession that he is a "homosexual"). Thus, in unusually lengthy sequences, the young men "tickle" each other on the cheek and lips (while denying they are gay) and, later and more comically, tackle the tricky mechanics of anal sex (the apparently more macho man bottoms once more).

Homosexuality is here at once innate (something one needs, in the words of the synopsis, to "self-accept") and acquired (learned as a practice or performance, even in its most intimate moments). And, unlike *Todo el mundo*, *Cuatro lunas* takes seriously the conflict between gay subjects and the society that surrounds them, even within what is once more a relatively tolerant metropolitan milieu. It is hardly surprising that the elderly Joaquín hides his bathhouse adventures from his doting wife and daughters, going so far as to use the money intended for his grandchildren's Christmas presents to pay the pricey prostitute. But even student Leo will insist on secrecy, lest he be singled out as a "joto" at the university (in fact the couple's straight friends actively encourage the relationship that he insists on denying). And, more gravely, the child Mauricio will be bullied at school after giving his slightly older cousin an (off screen) hand job that is apparently consensual. The established professionals Hugo and Andrés, however, are shown as having an accepting social life, although their experiences at work remain unestablished. Slow dancing and kissing at a party on a city rooftop, they embody a glamorous metropolitan lifestyle that will be torn apart not by social pressure but by Hugo's affair with another man.

An early sign of Hugo's betrayal of his longtime lover comes when he fails to turn up for a foreign movie with Andrés and his friends. In a scene similar to that in *Todo el mundo*, we see only the spectators' faces as we hear off screen dialogue, this time in French, not German. But, as mentioned earlier, cinephilia extends here, unlike in the lesbian movie, to the full range of audiovisual media. The boys discriminate finely between different role plays and editions of video games. Fito's mother, finally accepting, suggests to her son via an apparent commentary on her favorite telenovela that he is still in love with the closeted boyfriend whom he has broken up with. Once more, we see only the viewers' faces and hear off screen dialogue.

Cuatro lunas is relatively unchallenging in its narrative or technique, as the multistrand network narrative is familiar in Mexico since at least Jorge Fons's *El callejón de los milagros* ("Midaq Alley," 1995). And it takes care to offer gay viewers moments of unproblematic visual pleasure. The actors, often TV veterans, are consistently comely. And scenes in the sauna (shot at the traditionalist Baños Marbella in the old city center) and (later) a more intimidating red-lit sex club provide a chance for crowd-pleasing frontal male nudity. Unlike the asymmetrical framing and lengthy sequence shots favored by *Todo el mundo*, *Cuatro lunas*'s scenes generally break down quickly from two-shots that observe a couple's interaction to conventional shot reverse shots of good looking men's faces in tight close-up. These dialogue sequences encourage both fetishistic desire (for those handsome televisual males) and narcissistic identification (with those upwardly mobile middle-class citizens).

One touching scene halfway through the running time, just three minutes long, illustrates *Cuatro lunas*'s assured, mainstream technique and that technique's ideological implications. Twenty-something Fito has been invited to a party by his now long term boyfriend, the hitherto closeted Leo, where he will be introduced to Leo's family. The sequence begins with an unusually lengthy take of forty seconds, which sets up the tension. As Fito, smartly dressed in the yellow tie his mother has picked out for him, sits concerned in the foreground of his modest living room, his mother fixes a sandwich in the tiny kitchen behind. His lover is late and the mother suggests he call once more. Fito fakes a phone call, telling his mother that Leo has arrived to pick him up. As he exits we cut to a close-up of the mother's skeptical, concerned face. A plangent piano motif comes in on the soundtrack.

Next, down in the darkened street, we see Fito waiting anxiously. We cut to a high angle looking down on him from up in the apartment. Fito is suddenly small and vulnerable beneath the streetlamp. The camera cuts closer on Fito as his emotion increases. We cut back to the high angle. But this time it is revealed as a subjective shot from the mother's point of view. Back in the street Fito gets a text and we see, in turn, from his perspective an extreme close-up of his phone with a single word on its screen: "Perdóname" ("Forgive me"). Fito now returns to the apartment and his mother cradles her jilted, weeping son in her lap. Although both actors are awarded emotional close-ups that match the music that is still playing in the background, their two-shot is clearly reminiscent of a *pietà*, with the single mother a new Mary and the betrayed son a gay Christ.

Ironically enough, the pair is sitting on the couch where, earlier in the film, the then-homophobic mother had stonily watched her telenovela, refusing to hear her son's secret. Now her dialogue could not be more explicit: "You're too good for him. Another one will come along. I'll support you. I'll accept whoever you love, as long as he loves you. Anyone who doesn't love you isn't worthy of you." Like the mainstream shooting style, which clearly encourages us in traditional manner to identify with the two characters by adopting each of their viewpoints in turn, the dialogue serves the film's principal aim of social acceptance for gays and self-acceptance by gays. Homosexuality, once ostracized, is now incorporated into the heart of the family. And by the end of the film the lily-livered Leo will have presented Fito to his own parents also.

Avid viewers, whatever their age or social background, are thus taught, in spite of inevitable disappointments, both how to have an attractive gay love object and how to be a plausible gay subject. And while openly gay *Hollywood Reporter* critic Boyd van Hoeij complained (2014b) that the plotline of the older man would disappoint its audience by showing the persistence of the closet in modern Mexico, that story could be read as revealing how historically distinct modes of experiencing same-sex desire coexist in a single, complex present moment. Moreover, van Hoeij praises the professional polish of this low-budget drama, which he calls "slickly produced and well acted." These quality visuals underwrite *Cuatro lunas*'s aspirations to professional legitimacy, in cinema and beyond.

Such hallowed mechanisms of mainstream cinema were fed to and fostered by fans through social media. While *Cuatro lunas* lacked *Todo el mundo*'s institutional legitimation (it boasts no credits to government institutions), its director and producer are a real-life couple who publicly celebrated the anniversary of their own first "tickles," encouraging the audience to make the leap from their fiction to their life (such explicit coming out remains rare in Mexico). And they took care to build an audience long before the film's delayed theatrical release. That release came only a year after its premiere at the Guadalajara festival that had provided the project with development funding. The lengthy production process was also documented on Facebook by its homemade production company, Atko, which celebrated when it reached a thousand likes.

Special screenings and events (most recently around the publication of a novel based on the script) kept the film socially alive for its audience, after an extended commercial run with commercial exhibitor Cinépolis. This lasted an extraordinary twenty weeks and took in provincial cities around the country. Twitter feeds energetically promoting the film itself and its associated events included those featuring the often-shirtless actor Gustavo Egelhaaf (who had played the closeted Leo). His self-named "Egelfans" rallied to subsequent theater performances by their idol in Mexico City that were unrelated to his film role (the versatile Egelhaaf had begun his career as a TV presenter). Digital media were allied with physical events that enabled an experience of active fandom rare in Mexican cinema, especially in this vulnerable queer context, constructing a collectivity or multitude, if not quite a community.

Here the novel, a venture that is perhaps unique in Mexican indie film, is of the essence (Tovar Velarde 2015). As a print book, it provided fans with a tangible reminder of the ephemeral, albeit extended, theatrical screenings of the film (full disclosure: I first saw the film at the Morelia festival where it had received support under the Lab scheme, but, in the absence of a legal version, bought the DVD copy I used for this chapter from a pirate stall on a Mexico City street). And, within its fictional world, the novelization extends the scope of the movie plot to expand both social context and psychological motivation, enriching the intimacy of the fan's experience. Thus, most dramatically, we learn only in the book that aged Joaquín had suffered an unrequited love for a fellow student many years before (Tovar Velarde 2015, 97–99); and that his apparently clueless wife is fully aware and accepting of the gay desires her inexperienced husband barely recognizes himself (129). A small independent film thus cannily employed multimedia resources to extend its reach and influence. *Cuatro lunas* can thus be seen as a mainstream work that, unlike *Todo el mundo* (which has little social media presence), aspired not so much to achieve niche New Wave distinction as it did to attract as broad an audience as possible within necessary limits (financial, social) that it could not control.

What, then, of posthegemony? Although *Cuatro lunas* often shows male homosexuality to be fully integrated into modern Mexican society, still (unlike *Todo el mundo* once more) it testifies to the residue of homophobia. Indeed, in the introduction to his book (Tovar Velarde 2015, 7–13) the director bravely confesses to his own traumatic bullying at school as the origin of the cinematic story he wrote, shot, and cut so many years later. Yet, thinking perhaps of its target audience's wish fulfillment, the four stories end relatively happily. The boy is reconciled with his disapproving father, who teaches him how to defend himself against bullies. The students get back together when Leo comes out to his family, finally. And the professor enjoys a last family Christmas after a single, satisfying experience of gay sex (he bottoms for the straight-identified sex worker, who later attends the professor's homage and receives a volume of his poetry as a gift). Although the professional couple breaks up, they are shown at the film's conclusion to initiate brave new lives on their own. Andrés paints his new apartment a vibrant magenta. Hugo starts a solitary fridge magnet collection, now reconciled to the solaces of kitsch or camp that he had previously disavowed when they were enjoyed by his too effeminate boyfriend.

Todo el mundo's director said in an interview that he didn't make "gay films." *Cuatro lunas*'s cast and crew also dutifully repeated their film's universalizing tagline "Love is love." The limits of this discourse became apparent at the film's launch, however. Distinguished veteran Juan Manuel Bernal, who plays the homophobic father of the gay child, was questioned by the press as to his own rumored sexual preference after his supposed long-term relationship with a younger man had been outed in gossip magazine *TVNotas* (Martínez 2014). He replied that while he was committed as an actor to (unspecified) social movements, he preferred not to speak about a personal life that deserved to remain "private" (Cadena Tres Espectáculos 2015). This reticence had not prevented him, however, from playing a gay character in HBO Latin America's TV series *Capadocia* (2008–12), as we shall see in the next and final chapter.

In a similar way to the actor's insistence on personal privacy, what is wholly absent from the film is any sense of communal activism or public struggle based on a sexual preference. That preference is shown by *Cuatro lunas* itself to be multiple and protean in form but single and unchanging in nature, fundamental to a stable, socially accepted identity. The four lives (four moons) may be one, but the social transformation they embody (it embodies) in a formerly homophobic Mexico works on the audience, in posthegemonic style, through habit and affect, not political agency.

Trans Thriller: *Carmín tropical* ("Tropical Lipstick," Rigoberto Perezcano, 2014)

Es la historia de un regreso, el de Mabel a su pueblo de origen para hallar al asesino de su amiga Daniela. Un viaje por la nostalgia, el amor y la traición en un lugar donde el travestismo cobró en su día una inusual dimensión. (Conarte 2014)

This is the story of a return, that of Mabel to her hometown, in order to find the killer of her friend Daniela. It is a journey through nostalgia, love, and betrayal in a place where transvestism [*sic*] took on at the time an unusual dimension.

In an interview with journalist Arturo Aguilar (2015) in the online media magazine *Gatopardo* (the title is clearly cinephile), director Rigoberto Perezcano calls attention to the transgeneric nature of his transgender film, his second feature. Thus, he says, *Carmín tropical* begins as a documentary, continues as a drama ("ficción"), and ends as a thriller or film noir. In cinephile style, then, his filmmaking project is reflexive. Just as his debut film, the well-received *Norteado* ("Northless," 2009), treated the theme of migration to the United States in a new, comedic style far from the normal social realism, so this feature, the only one in world cinema he claims in which a "travesti" investigates a murder, reworks the traditional conventions of films made either for the festival circuit or the commercial marketplace.

It is striking also that the one publicity shot reproduced by the press in reviews of *Carmín tropical*, which is in fact uncharacteristic of the film, also cites a self-conscious tradition of movie history. The still comes from the climactic ending of the film where

Perezcano shows, for the first and only time, his heroine Mabel in full theatrical costume and makeup as she performs a torch song on stage. In a Q&A that I heard after the film's world premiere at the Morelia festival, actor José Pescina, dressed with male sobriety, said he had used cinematic icons from classic Hollywood and the Mexican Golden Age (Joan Crawford and María Félix) as models for how to walk and talk as a woman. The fact that the film's director received a Guggenheim award when he was developing the script for his film also suggests a US connection that might perhaps favor a more mainstream project. The minimalist Mexican features more typical of those shown at Morelia, on the other hand, tend to be funded by experimental European festivals such as Rotterdam.

In spite of this cinematic citationality, *Carmín tropical* is, however, as its director suggested, grounded in a physical location and one that is far indeed from the somewhat deracinated metropolitan milieu of *Todo el mundo* and *Cuatro lunas*. The film is set quite precisely in Juchitán, a coastal town on an isthmus of Tehuantepec in the director's native state of Oaxaca, which is well known for hosting a distinct subculture: a third sex of *muxes* (transgender women) irreducible to international models of homosexuality and integrated into traditional society. Hence, the contemporary documentary intent of a film that seeks, nonetheless, to revive an archaic cinematic genre, the film noir that was invented by Parisian cinephiles so long ago.

Perezcano takes care to make clear in interview that, although his main characters are played by actors from outside Oaxaca, the supporting cast are locals who not only appeared on screen but also served as specialist consultants in order to ensure the authenticity of the professionals' performances. Unlike in *Cuatro lunas*, however, there are no comely TV or theater stars here to encourage fan participation in a world far from modern Mexico. *Carmín tropical* would, however, figure prominently in critics' lists of the best Mexican movies of the year.

Documentary intent is revealed by cinematic technique in *Carmín tropical*'s opening sequences, even as the film testifies also to a transmedia focus typical of my Mexican mainstream films. In this case, transmedia interest is focused on still photography. Thus, after credits that reveal that the film is funded in part by the Oaxacan state authorities (who clearly see no problem in promoting the attractions of their *muxe* subculture), we are shown the family snapshots in which a fragile young boy grows up into a self-assured and glamorous woman who will prove to be a murder victim. Perezcano cuts without explanation to an alienating environment, the deafening factory where his protagonist Mabel is working. We next see her packing a modest bag and traveling cross-country by bus, the cause of her journey still unknown. Our disorientation is not complete, however, as we are given an explanatory voiceover from the main character.

A further technique is crucial here and throughout the film. *Carmín tropical*'s camera (operated by Alejandro Cantú, the veteran of Julián Hernández's features) sticks close to the film's determined heroine as she walks in high heels through the authentic Oaxacan locations (handsome beaches, less picturesque unmade roads) holding tight to the back of her head. It is a shooting style that might suggest both casual documentary and film noir pursuit. Within the fiction, Mabel is indeed being followed and will, finally, be the next victim of the transphobic killer.

In the main body of the film, however, Perezcano takes care to show the integration of *muxe* identity into all of the institutions of this regional society. Mabel sits preparing food with members of the family of her dead friend, who confess that the victim was the "favorite" of her parents and, as the sole "daughter," was expected to care for her mother and father in their old age. In the course of her investigation Mabel visits a prison where a suspect (the victim's ex-boyfriend) is being held. As she enters a prison she is asked for her sex. When she replies "muxe," the official writes it down without turning a hair. And the straight-acting prisoner visited by Mabel insists that loving a *muxe* did not make him any less of a man. Mabel and her dead friend are shown to share a group of supportive friends, some of whom seem to be cis-gender gay men. And they work in a local nightclub that appears to cater to a wide variety of clients. Some *muxes* are strippers there, others (like Mabel) only chanteuses.

It is significant that locals, apparently at ease in their home environment, struggle to remember the term to describe someone who hates "people like us" (Mabel tells them it is "homophobe"). So taken for granted is the *muxe* identity in a Oaxacan context that is well known in the rest of Mexico, that Perezcano never takes the trouble to investigate or explore its psychic or somatic specificities. For example, we are not told if *Carmín tropical*'s convincingly feminine trans women have had recourse to hormones or gender confirmation surgery.

Clearly the anomaly in this accepting context is the murder. Where straight-themed festival films in Mexico (such as Amat Escalante's *Heli* of 2013) present fatal violence as casually commonplace, especially in a criminal context, here a single killing is shown to have continuing and devastating effects on a tightly bound law-abiding community. And, as mentioned earlier, Perezcano's film moves slowly from documentary via drama to thriller. Casually introducing the taxi driver from out of town who will drive Mabel to the prison and nightclub and is deceptively called "Modesto," Perezcano introduces a touching romance into his murder mystery premise. The cautious Mabel slowly succumbs to the unshowy charms of this straight-acting man, a lover of romantic songs, who treats her always with respect. One excursion to the beach is especially idyllic. The noir revelation comes, however, once more with a gift. A key piece of evidence in Mabel's investigation is a photo of her dead friend in which a man's face, presumably that of the assassin, has been cut out of the picture. At the climax (which is crosscut with Mabel's musical performance) the too-good-to-be-true taxi driver presents her with a gift box. She takes off the lid to reveal the missing fragment of the photo. As we cut to black, we learn (she learns) that her perfect lover is the murderer.

This shocking twist in the tale is, of course, a highly commercial technique typical of the murder mystery. Called by *Hollywood Reporter* critic Boyd van Hoeij (2014a) once more "a killer ending," it could hardly be further from the casual plotting of auteurist or festival films. These include Perezcano's own earlier *Norteado*, which is the meandering story of a wanderer as he seeks, slowly, to cross the border to the United States. Perezcano thus signals the mainstream status of his second film by securing its allegiance to film noir. It is a technique of cinephilic citation founded on a classically Hitchcockian McGuffin (here the missing photo fragment).

More important, however, is the director's posthegemonic vision of Mexico. Here in Oaxaca at least (and Perezcano, a self-identified indigenous Zapotec, is explicit in his desire to represent his neglected home region in his films) there is neither repression nor resistance. This is a community where queer life (the everyday habits of the accepting family) and queer performance (the warm affect of the welcoming nightclub and circle of friends) are their own justification. In spite of the communal focus, however, Mabel's investigation and subsequent death are solitary and individual. And the murder of her friend had not given rise to a collective, popular initiative from the grieving community of Oaxacans and *muxes*. Nonetheless, however, *Carmín tropical*'s final and unforced moral is that it is transphobia not transgender that is the anomaly. The isolated straight-acting killer, a stranger from a strange town, is the only aberration, and one whose evident psychopathy remains inexplicable, uncanny within the world of the film.

A Gift of Love

I saw *Cuatro lunas*, as mentioned earlier, first at a Morelia screening a year after its premiere and then on an expertly English-subtitled pirate DVD purchased on the street in Mexico City (the vendor disappeared in search of the film and then ran after me with his precious discovery). *Carmín tropical* I watched also at Morelia, where it won best Mexican fiction feature, a rare achievement in a festival normally known for its support of minimalist or contemplative art cinema. *Todo el mundo* proved more elusive. I finally saw it in a Mexico City hotel room with my iPad connected to a generous friend's local Netflix account. At the time of writing, *Cuatro lunas* is, however, available on Netflix in the United States, unlike the other two films. It even received a brief New York theatrical outing a full year before it reached theaters in its home country. Tovar Velarde and Perezcano are also accessible on Twitter, while the director of *Todo el mundo*, Raúl Fuentes, keeps a dignified social media silence.

Although I call these films "mainstream," they are variably accessible in their distribution and exhibition. And, aesthetically, they have been disadvantaged in their reception by falling between the formats of art cinema and popular movies. Thus, festival films from Mexico (as elsewhere) tend to take the form of slow cinema, and, as mentioned elsewhere in this book, there has been a strong trend for artistic minimalism laced with sex and violence (Carlos Reygadas and Amat Escalante are exemplary here). It is telling that, following the US and European demand for bleak social realism from Latin America, the only gay-themed film from the region recently to win major awards is one based on a theme unmentioned in my mainstream triad, street prostitution and violent crime: Venezuelan Lorenzo Vigas's *Desde allá* ("From Afar") won the Golden Lion in Venice in 2015.

The theme of sex work in *Cuatro lunas*, on the other hand, is characteristically sweetened or softened by the mutual offering of gifts: when the professor presents his book at the end of the film, the younger man proffers an unexpected (and unpaid) kiss. Conversely, the revelation in *Todo el mundo* and *Cuatro lunas* that a large Mexican middle class exists would no doubt prove unwelcome to First World cinephile audiences who prefer to believe that the more explicit the evidence of poverty and violence is in a Mexican film, the truer that film is to the experience of that country. Even the rural community of *Carmín tropical* seems relatively comfortable, revealing no evidence of the misery and brutality typical of the social realist features from Latin America that have, since at least Buñuel's *Los olvidados* (1950), monopolized foreign views of the region.

While consciously or not (and Perezcano for one is very conscious of this) the films in this chapter avoid formats favored for foreign festivals, they also stray from the templates for domestic box office hits. During this period there was a trend in Mexico for hugely successful heterosexual romantic comedies (rarely seen abroad) that lacked even the token presence of a stereotype from the genre's glory days in the United States in the 1990s: the bride's best gay friend (TV star Eugenio Derbez's *No se aceptan devoluciones* ["Instructions Not Included," 2013] is a partial exception here, as his straight male protagonist's former partner is granted a lesbian lover who is not the object of ridicule). The social reality of a newly middle-income country with a relatively accepted LGBT community is thus depicted only in my mainstream films, perilously poised as they are between art and commerce, and this is one of their great virtues.

We have seen, also, that while these films may invoke cinephile references with varying degrees of reflexivity (*Todo el mundo* is the artiest and archest, *Cuatro lunas* the most earnest), they also seek, varyingly once more, to connect with or even create a collectivity of fans. Two years after its festival premiere, *Cuatro lunas*, the most successful in this area, still has some eight thousand followers on Twitter, proof of its enduring link to an audience that continues to turn out for physical events like the launch of the book of the film. Conversely *Todo el mundo* may not have been helped in its reception by having a male director, even though he voiced the universalist party line that his was not a lesbian film or (in *Cuatro lunas*'s words) that love is, quite simply, love.

Ironically, perhaps, the films themselves add a gloss to the dominant discourse of queer visibility in Mexico, that of "diversity." Lesbian, gay male, and transgender lives are shown in each respective feature to be wholly separate from each other, rarely if ever intersecting. This isolation might perhaps be related to a posthegemonic era in which collective alliances against common enemies prove difficult to realize or even to imagine.

But, as I suggested in my introduction to this chapter, the value of these films is that they suggest that queerness, understood in all its variety and irreducibility, could be a new kind of consensus for a modern Mexico eager to show off its recently acquired social liberalism, at least in the middle-class milieu that two of my features depict. *Carmín tropical*'s *muxe* environment is an outlier here, but one that adds an invaluable regional perspective to the other films' metropolitan setting. Finally, then, all three films serve as a gift of love to the emerging, fragile collective of a queer-interested audience between the art house and the multiplex, in Mexico and beyond, a gift that transcends both the symbolic exchange of the international festival circuit and the more mercantile economy of the national commercial exhibitors. We can see now and finally how quality television on LGBT themes also aspires to this position in between art and commerce.

ARGOS'S QUEER TELENOVELA

I am watching TV in my Mexico City hotel room on a Sunday afternoon in March 2016 and feeling a sense of déjà vu. A comedy called *Vecinos* ("Neighbors") is being shown on Channel 2, better known as Televisa's "Channel of the Stars," still Mexico's highest-rated station. Set in a small, rundown apartment building, *Vecinos* focuses on the everyday problems of urban life among a group of residents who include such stereotypes as the washed-up actor, the old maid, the hen-pecked husband, and the deadbeat doorman with a comic regional accent.

In this episode, a rumor spreads among the neighbors that two men who live together in the same apartment are gay. A retired military man says he won't stand for that sort of thing in his building. And the group sends up a young woman to communicate the community's displeasure to them. But the rumor soon turns out to be false. The couple are unimpeachably heterosexual friends whose secret is just that they have been illicitly selling sportswear out of their home. And the reason a gaggle of effeminate young men have been making their way upstairs to their apartment ("Two for the price of one!" they exclaim, to the horror of the neighbors) is that the garments are embellished with a rainbow flag, the meaning of which has eluded the clueless vendors. Imagine their embarrassment when they find out the cause of their sudden popularity with *jotos* ("fags")!

The reason for my sense of eerie familiarity with the show was that I recognized it as a Mexican remake of a Spanish original, *Aquí no hay quien viva* ("No One Can Live Here," Antena 3, 2003–6). It is a series that has seen many versions in Latin America and beyond (in France it was called *Faites comme chez vous* or "Make Yourself at Home"). But, as far as I know, it is only in Mexico that the scriptwriters take care to specify that the male couple in the building are not gay. Indeed, in the very first Spanish episode (broadcast as long ago as September 7, 2003), nosy neighbors also intruded on the home of a male couple, who resented their interference but were happy to acknowledge their homosexual preference. The fact that *Vecinos* is produced by Eugenio Derbez, the most powerful figure in Mexican TV and film comedy, who surely had some discretion over his adaptation of the franchise, suggests that this erasure of homosexuality runs deep in the culture of Mexican television.

Although at the time of writing *Vecinos* is being rerun on Televisa's main channel in a block of four episodes every Sunday afternoon, the series is ten years old: new episodes aired only from 2005 to 2007 (this one was first shown on September 7, 2005). Yet it would be a mistake to think that *Vecinos* remains uncharacteristic of Mexican TV fiction since that time. In 2010 Álvaro Cueva, the most distinguished TV critic in the country who writes an excellent daily column in *Milenio*, published a piece called "Tele marica" ("Queer TV") that traced a devastating panorama. Cueva begins by saying that Mexican television and homosexuality are "incompatible" because *jotos* and lesbians "don't count and don't bring in any money" (Cueva 2010, 176). As sexual orientation is not measured as a variable for ratings purposes, TV companies are never going to pay attention to LGBT people (177). Moreover, the major advertisers are few in Mexico and contract out their campaigns to agencies that are "100% conservative" (178).

Turning to content, Cueva writes that surveying the history of homosexuality on Mexican TV is a thankless task because, although there has been progress in society over the years, the small screen seems to belong "to another planet rather than to our country" (Cueva 2010, 180). Unlike the United States, Mexico never had on its dominant networks a *Dynasty* or *Melrose Place* back in the 1980s and 1990s or a *Friends* or *Will & Grace* going into the 2000s. Even the relatively more progressive Spanish-language telenovelas made in Miami feature only clichéd gay best friends for their heroines (181). Mexico suffered rather the "grotesque figures" of queens on comedies and talk shows (182). (The truly appalling *Hora pico* ["Rush Hour," 2000–2007] sketch show, with its stereotypical waiter character, was also being rerun in 2016 on Televisa's heritage channel XEQ Gala.) So bad is the situation, writes Cueva, that LGBT people no longer hope for better characters or more decent shows on Mexican TV. They are resigned to letting the TV companies do what they want and have moved on to pay television or the Internet for entertainment.

Cueva makes just one exception in his negative prognosis:

> La cúspide de la homosexualidad vista en televisión fue el personaje que José María Yazpic [*sic*, for Yazpik] interpretó en *La vida en el espejo* ya que el muchacho asumía con firmeza su homosexualidad y lejos de parecer una "loca," era un chico viril, común y corriente, lo cual desconcertaba a las buenas conciencias. (Cueva 2010, 181)

> The high point of homosexuality ever seen on television was the character played by José María Yazpik on *Life in the Mirror*, since the young man was resolved to come to terms with his homosexuality and, far from looking like a "queen," was a masculine-acting youth who was perfectly ordinary, all of which unsettled conventional morality.

There is thus a possibility of small but honorable exceptions to Cueva's negative vision of "TV marica" in Mexico. And the outstanding *La vida en el espejo* (screened by mainstream broadcaster Azteca but made by independent producer Argos) will be the first text I treat in this chapter on LGBT TV.

If we look more broadly at the history and character of Mexican television fiction, Argos and *La vida en el espejo* keep cropping up. This is true of three essays with very different approaches published in a 2011 volume called tellingly *Telenovelas en México: Nuestras íntimas extrañas* ("Telenovelas in Mexico: Our Intimate Strangers") (perhaps gays are also "intimate strangers," always present but ever ignored on the Mexican TV scene). Álvaro Cueva once more gives a negative panorama in his "confessions of a telenovela critic" (Cueva 2011, 13–44). He writes that where once Mexico was the greatest world force in the genre, now other countries are way in front (13). Mexican shows do not allow their heroines actively to pursue men and promote "the culture of suffering" (of submission, resignation, and trusting to the Virgin), rather than reliance on education or even physical exercise (15). Likewise, in Mexican melodramas, wealth is never achieved by hard work, but only by means of a lucky inheritance or fairy-tale marriage (17). Too many current telenovelas, Cueva writes, are either remakes of old titles (which no longer correspond to social reality in Mexico) or adaptations of foreign formats (which do not coincide with Mexican cultural characteristics) (18–19). Casts are of poor quality, as good actors now prefer to work elsewhere (24). Standards in music and costume are lower still (25–26).

Showing the author's ambivalence to the medium to which he has dedicated his professional life, the second half of Cueva's essay rehearses all the points above, but with a more positive spin (for example, he now writes that it takes real skill to adapt a foreign title to the domestic market). But in both halves of the article Cueva signals out one producer as an exception who succeeds in making quality content without succumbing to the pressure of brute market forces: Epigmenio Ibarra, the CEO of Argos (Cueva 2011, 25, 40).

In the same volume, Rubén Jara, a founder of the media statistical institute IBOPE, offers a very different survey of Mexican TV, focusing this time on ratings (Jara 2011, 125–60). Telenovela, he writes, is like "green [chili] sauce," ubiquitous in Mexican households (125). In 2010 alone the genre extended over 128,000 minutes and 3,800 episodes. Yet the audience profile skews more female, elderly, and impoverished (129) than the general population, with 42 percent of higher social classes (ABC1) preferring to watch subscription cable services rather than the free to air programming of Televisa and Azteca (130).

The telenovela format has, nonetheless, proved resilient, shrugging off a "fleeting" challenge from reality shows (Jara 2011, 135) and continuing to attract a "family audience" that no longer exists in other territories, where networks now tend to seek a quality demographic. With more babies and pregnancies than the population as a whole, it is perhaps unsurprising that this "telenovelero" public is also more Catholic and more attached to traditional gender norms than is typical of modern Mexico (136). Every weekday prime time "between 7 [actually 7.30] and 10.30 pm" telenovela remains dominant for such audiences (138). So ubiquitous is it that Jara calls the genre "another member of the family" (158).

Contradicting this pervasive social conservativism once more are Argos's productions, which, under these immobilist circumstances, have proved surprisingly popular. *Mirada de mujer* ("A Woman's Look," 1997), a feminist slanted look at adultery, was the highest rating show on Azteca in the entire twelve-year period from 1998 to 2010. *La vida en el espejo*, where Yazpik's character famously said, "Dad, I'm homosexual," also figures strongly, appearing at number six in the top ten (143). Like Argos's more recent overtly feminist and lesbian-themed telenovela *Las Aparicio* (Cadena 3, 2010–11), which I have studied elsewhere (Smith 2014, 137–51), such titles attracted valuable audiences of higher social class and greater youth than the free to air average (146). Even in Jara's strictly commercial terms of ratings, there would seem to be a financial case for innovation and quality over business as usual for a *telenovelero* audience of aging, financially stressed, and undereducated Catholic housewives.

Finally, Guillermo Orozco, Mexico's most esteemed academic TV commentator and the current director of OBITEL, an international survey of Spanish-language series, gives his own account of the field (Orozco 2011, 185–218). He notes that the audience for telenovelas has recently fallen (185), although current ratings are still high enough to attract advertisers (186). Like Cueva, he calls attention to the anachronism of modern titles, obsessed with themes such as virginity (188) or with characters such as *Simplemente María* or "simply María" (this tale of an innocent young girl from the country, originally produced in 1989, would surface in a remake on Televisa as late as 2016) (189). Citing Raymond Williams, Orozco calls such out-of-date cultural phenomena "residual," holdovers from another era (189).

Yet in the past, Orozco writes, telenovela did carry a social message, taking care to inform viewers in the 1970s and 1980s on such topics as contraception, literacy, and public health (Orozco 2011, 201–2). And since 2009, he claims, there has been a decline in the typically conservative Televisa-style series: atemporal, naively romantic, and devoid of reference to national context (204). Hence, even now there is a chance that telenovela can help to set the "social agenda" (216). Yet, while the government has no direct influence on TV programming, it can clearly influence it in negative ways through its massive advertising budget.

For Orozco, as for Cueva and Jara, if we look for a "new telenovela" we should go back to Argos and its first innovative success *Mirada de mujer* (Orozco 2011, 211). And I would suggest that, in spite of the alleged ephemerality of television, viewers have long memories. They thus have no difficulty in retaining the social impact of significant past programming that manages to attain the status of event. In the case of LGBT TV, the condition of domestic consumption (the idea that telenovela is as common in Mexican households as green sauce) surely heightens the intensity of that impact when challenging new series are welcomed into the home (or not, as the case may be) as a fellow member of the family.

While Mexicans visit movie theaters just three times each year (and Hollywood fare is dominant in the films on offer), they watch four hours and forty-five minutes of overwhelmingly local TV each day (up by thirty minutes on twelve years earlier) (Jara 2011, 128). There can be little doubt, then, as to which medium makes the greater impression on audiences and forms a more substantial part of everyday life. And although the cinephiles who seek out Julián Hernández's features in art houses would no doubt deny that they watch television, the more mainstream viewers of the films I studied in the last chapter such as *Cuatro lunas* would probably take pleasure in seeing their favorite actors habitually on television, as well as on rare occasions on the big screen.

Evidence for this lasting impact of television comes from the unlikely source of glossy monthly lifestyle magazine *Capital55* (the number in the name is taken from Mexico City's area code). The cover of the April 2015 edition shows the actor who played the gay youth in *La vida en el espejo* sixteen years after his first great success. A now handsomely bearded and purposeful José María Yazpik poses on the cover in a gray suit and white shirt, both by British luxury brand Burberry, accompanied by the strapline "Sin rodeos" ("Plain Talking") (Rodríguez 2015). With its subject said to be enjoying his "artistic and personal maturity" (48), the article begins by rehearsing Yazpik's impressive curriculum in theater and cinema, including a comedy for Almodóvar and, most recently, an austere Mexican art movie, *Las oscuras primaveras* ("The Dark Springs") (Rodríguez 2015, 49). It would appear that playing a gay character did not by any means harm his career, even though he did so back in a less tolerant 1999 and when he was virtually unknown.

Yazpik's first successful role, then, still fondly remembered, was in the Argos telenovela. The journalist asks if it was difficult preparing to play a "homosexual" when Mexico wasn't as open to such themes as it is now. Yazpik replies that it was difficult not for him but for those who "didn't want to hear" what the series was saying. And he claims that his character Mauricio was the first in the country to be openly gay and not depicted as a caricature. But if there was any tension on set it was more than matched by controversy off: one advertiser, a car brand, quit the show because it didn't want to be associated with gays. Yazpik's only concern, however, was to construct a "credible character," a process for which he asked his gay friends how they had experienced coming out of the closet and dealing with their family and friends. Although he subsequently made several more telenovelas, he says he would never do another, blaming their characteristically slow rhythm: while taping the last one, he says, he spent months acting the same scene over and over again, changing only his tie (Rodríguez 2015, 51).

I will argue later that this slowness can be a vital technique in integrating fictional characters into the existence of real life viewers who share with their favorite telenovelas long months of daily cohabitation. Moreover, in addition to the more memorable characteristics of novelty and innovation, the neglected qualities of familiarity and repetition can also be important virtues in long form television, cementing as they do the parasocial closeness of audience and characters. But what the *Capital55* interview reveals is that the quality series made by Argos managed to make a distinct and lasting impression in the huge flow of Mexican telenovela. And that a single scene of televisual coming out (a theme that is, as we have seen in previous chapters, rare indeed in Mexican LGBT cinema) can still be expected to attract the interest of upscale readers of a lifestyle magazine almost twenty years after they first saw it.

In the rest of this last chapter, I will discuss three titles by Argos, making an implicit argument for the auteurship of the independent production company in the television medium, as distinct from that of the director in art cinema. The fact that CEO Epigmenio Ibarra came to fame as a Leftist journalist with a close relationship to the Zapatistas (a rebel militia in 1990s Chiapas) no doubt still influences his titles' political bias. Also skilled in social media, the vituperative Ibarra is, uniquely for a TV producer in Mexico, a major public figure who currently has over a quarter of a million followers on Twitter.

Yet, as we shall see, each series is textually different from the other. And their different production date enables us to plot the changes in depiction of LGBT characters over the course of two decades. Moreover, reflecting the complexities of television production and authorship, each was made by Argos to be shown on different channels with which the producer collaborated: from free to air network Azteca (*La vida en el espejo*, 1999) to radical upstart Cadena 3 (*El sexo débil*, 2011) and boundary-pushing subscription channel HBO Latin America (*Capadocia*, 2008–12).

Family Plots: *La vida en el espejo* ("Life in the Mirror," Azteca, 1999)

Santiago and Isabel have been married for 25 years. She is a successful public relations executive and he looks after the three children at home. At work Isabel has met Eduardo, a handsome young man with whom she is having a passionate affair. Julio, a close friend of the family, will tell Santiago everything. Santiago tries to save the situation, but when they see this is impossible both decide to get a divorce. Gabriela, a young radio announcer, meets Santiago and they fall in love with each other. Meanwhile Isabel starts to regret her affair. She tries to leave Eduardo but he goes crazy and tries to kill her.

In the subplots we see the lives of the children of Santiago and Isabel. The eldest son, Mauricio, seems to love his fiancée, but is impotent when he tries to have sex with her. On a trip with one of his best male friends he discovers his true sexual identity. Diana is a teenager suffering from the problems common to adolescents. The youngest Eugenio gets his girlfriend pregnant and must face up to being a father at an early age.

In this synopsis, the gay theme, in which I am especially interested, seems subordinate. But the summary shows that it is embedded in and emerges from a rich and complex narrative and social context in which the demands of family merge with the worlds of work and sociability. Moreover, it also suggests that the appearance for the first time in a telenovela of a credible homosexual character depends on the breakdown of a sex/gender system that, from the very start, is seen no longer to function for modern Mexico.

Yet the series' credits are more ambiguous. All of the many characters appear in shimmering images that fold and fade into one another, suggesting the mirror of the title. And the plangent lyrics of the theme song, performed by Spanish married couple Ana Belén and Víctor Manuel and set to a sultry saxophone accompaniment, are also somewhat ambiguous. One partner, we are told, gives the other what she has too much of, while another offers her what she lacks, suggesting a lack of reciprocity even in the exchanging of gifts. Finally, Belén sings, "through mirrors / my soul slips away from me" ("por los espejos / se me escapa el alma"). It is a resonant image of subjective loss or dissolution that will prove characteristic of this new telenovela for troubled times.

The Leftist daily *La Jornada* gives us the clearest and fullest account of the production process of this new venture in quality TV written on the occasion of the press launch for *La vida en el espejo* (García Hernández 1999). It is an apparently unlikely source given Mexican intellectuals' profound hostility to television, and a sign in itself that the series is something other than business as usual. Indeed, the article stresses from the start the novelty of this show, revealing as it does the "flip side of the coin" from the conventional female melodrama: the masculine world of "dreams, sorrows, fears, strengths, and weaknesses." It goes on to cite the track record of chief writer Bernardo Romero and director Antonio Serrano, who have given Azteca its biggest rating successes, such as the female-centered *Mirada de mujer* to which this might be seen as a sequel. While Romero stresses that "tears" are not the sole privilege of women, the two stars likewise emphasize the new complexity of their characters: Gonzalo Vega says that his abandoned father and househusband Santiago is not "lukewarm" but truly passionate about his wife, and Rebecca Jones that the adulterous mother and working woman Isabel truly loves her career "and men" (in the plural). The couple cannot be reduced, she says, to the old stereotypes of cuckold and whore.

For his part Epigmenio Ibarra states, on behalf of producer Argos, that his company is dedicated to producing "quality television with high production values and writers that respect the audience," even as the show conveys the "passion, feelings, and emotion" essential to telenovela. *La Jornada* readers were no doubt reassured to read, moreover, that part of this "quality" programming is "making fun of the genre itself" and parodying "entertainment journalism and its tendency toward gossip and sensationalism." Finally, director Serrano, who concurrently had romantic comedy *Sexo, amor y lágrimas* ("Sex, Shame, and Tears," 1999) playing in movie theaters (there will be more than one reference to the title of the film in the telenovela itself), claimed that, in spite of all the factors that "escape his control," he tries to "impose a personal style" on the telenovelas he directs. While TV authorship is collective (and *La vida en el espejo* has more in common with previous Argos TV titles than with Serrano's few movie features), the series will, as we shall see, demonstrate stylistic ambitions that are rare indeed in its format.

The article makes no reference to the daring theme of homosexuality and does not cite Yazpik's name. It is not clear whether this is to avoid spoilers (Mauricio will not come out until a long way into the over one hundred episodes) or to avoid alienating conservative viewers, otherwise attracted to the telenovela format. And if we look at the first weeks of the show itself (five prime-time episodes were broadcast from Monday to Friday) we see immediately the complexity of a fictional world and a televisual style that, even before the gay theme becomes explicit, consistently contradict the Televisa-style conservative model of telenovela outlined by the three sources in my introduction to this chapter.

This divergence from the norm is perhaps not surprising given the production context hinted at above. First, Azteca, then a relatively new rival to established giant Televisa, sought at that time to distinguish itself from its rival, even hiring Rebecca Jones, who claimed to be impatient with playing traditional roles, away from the dominant broadcaster. Second, when I interviewed Leticia López Margalli, one of the show's screenwriters, she told me that at the audition process Argos sought out young writers, like she was at the time, who had never worked before in television (Smith 2014, 250).

Further evidence for *La vida en el espejo*'s quality is that soon after the series ended, its actors went on to star in prestige feature films. Diego Luna, who plays Mauricio's younger brattish brother Eugenio, famously featured as another spoiled rich kid in *Y tu mamá también* (Alfonso Cuarón, 2001). Àlvaro Guerrero (Santiago's best friend) appeared in a further celebrated story of adultery (*Amores perros*, Alejandro Gutiérrez Iñárritu, 2000). Mauricio's neurotic sister Diana (Edwarda Gurrola) has enjoyed a notable career in indie film and television. And we have seen that José María Yazpik established himself as one of Mexico's best-known and most respected actors in all media.

How then do the opening episodes reveal the twin obsolescence of the sex/gender system and its favored genre, the telenovela, in modern Mexico? The main setting, the family's palatial home complete with extensive garden and loyal retinue of servants, is indeed lavish, as is typical of telenovela. But it is established early on that this home was not inherited but rather earned by the hard work of the married couple who set up a successful public relations firm together (thus breaking the number one rule of the workplace in telenovela). But by providing a lush costume and set design, *La vida en el espejo* continues to offer old-fashioned viewers the visual pleasure they associate with their favorite genre, even as this new kind of drama breaks the rules of traditional narrative.

The opening episode stages a radical challenge to the gender norms of telenovela. Confident Isabel, in a pantsuit, is first shown holding down the fort at work as caring Santiago copes with the crises of the couple's three squabbling children at home. We soon learn that Isabel is having a passionate affair with Eduardo, a young colleague who has fallen in love with her but whom she simply uses for sex (we remember that active female desire, like well-remunerated hard work, was impossible in the Televisa-style telenovela). When Isabel finally makes it home (youngest son Eugenio claims he hasn't seen her for a week), she is soon making secret sexy calls to her lover on her phone. And she arranges to go to Acapulco with him for the weekend, allegedly for work. All this in the first episode.

What is striking here is the defense of a mature woman's sexual appetite, even in the most unsympathetic circumstances. Before the series even begins, she has betrayed her doting househusband, who is far indeed from the brutal Mexican macho of the past. And she (and the series) treat her younger lover, who is also her employee, as a sex object: while he is frequently shirtless and occasionally fully naked (from behind), she keeps covered up, albeit in a lacy negligee, even in bed. One of the couple's female friends defends Isabel's behavior on feminist grounds: no one in Mexico would turn a hair if a middle-aged man took a much younger female lover. And the real problem, the friend says, is that "in this country" it is thought improper for a wife to be more successful at work than her husband.

Given these changes in what counts as acceptable female behavior, it is unsurprising that the series should investigate from the start the crisis in masculinity that its producers said was its distinctive theme. The second episode sees Santiago complaining of the failing body of a man in his fifties (he will often inspect his aging face in the mirror) and pleading for more free time with his wife even as she secretly reads a letter from her lover. Santiago's unsympathetic unmarried sister warns him that it is Isabel who is "the man in the household."

Yet Santiago himself has long been fashioning a softer, more emotional manhood. He fondly remembers how he raised his children, even explaining about menstruation to his daughter and confiding to his son that: "It's a lie that men don't cry." Typically, when Santiago learns of his wife's infidelity (episode 6), rather than violently confronting her, he will pretend he knows nothing, ordering in her favorite French food as a romantic gesture. The lavish banquet will, however, grow cold as she heedlessly eats Chinese takeout in bed at her lover's apartment. Santiago's deep disturbance is represented in a rare antirealist moment in episode 7: in a crowded shopping mall, he sees that all of his fellow shoppers are frozen in place, unable to move. It is a joint critique of bourgeois marriage and of capitalist consumerism that is no doubt in tune with the sensibilities of both the Leftist producers of the series and the readers of daily *La Jornada*, which reported on its premiere.

Beyond such seriousness (and the series is well aware that melancholy is as central to telenovela as visual pleasure), *La vida en el espejo* also parodies the new sexual setup. Santiago has a divorced friend called Julio who is "the King of the Machos" and spars playfully about feminism with young radio journalist Gabriela (Santiago's future lover). As the central couple stagger toward divorce, family members repeatedly attack each other for acting "as if they were in a telenovela," part of Argos's satirical approach to the conventions of their chosen genre.

But it is Gabriela's radio show that is the main tool used to explore modern media. The young radio presenter resists her management's instructions that women want to hear only about celebrity gossip and seeks out political issues to report on (the first is a public event intended to celebrate women organized by Isabel which turns out to be a failure). Gabriela even discusses the real-life forthcoming general election, which would lead to the end of the PRI era, arguing that voters should favor only candidates who support more women in government. Her ethical and political bias in radio is clearly analogous to Argos's position in television. It is a narrative strategy that allows *La vida en el espejo* to acquire cultural capital by distancing itself from the traditional telenovela that provides the stars for the gossip shows and magazines the show critiques as fake.

All pervasive in the series is a sense of pretense. Isabel is, after all, a specialist in public relations (she proves less successful with the private variety) and is baptized by her now wary husband as a "professional of the smile" (episode 7). In this world of simulacra, authenticity comes to be problematic. Characters explicitly recognize and call attention to the psychic causes of their own desires. Fiftyish Gonzalo is a father-substitute for the orphaned Gabriela. Conversely, teenage Eugenio's "substitute mother" is a loyal maid. Diana cites her brother Mauricio's closeness to their mother as evidence for an Oedipus complex, long before he comes out even to himself (episode 9).

In a media-saturated society, style takes over from substance, identification from identity. Conversely, however, the agonizing slowness, increasing familiarity, and insistent repetition of *La vida en el espejo*'s narrative enables mediated desires to be incorporated into the everyday life of viewers. The great numbers of such spectators suggest that they were fascinated by this newly complex mode of serial melodrama. The overt stylization of the shooting, to which I return later, must also have engaged viewers in new ways, alternately involving them in or distancing them from the fictional world. To take just one recurrent example, family members are frequently shot framed by doors and windows, pathetic prisoners of their life of luxury.

Having established this familial and social context, we can now turn to the story of Mauricio, the first complex and sympathetic gay character in the history of Mexican television. Current viewers can see an accelerated and concentrated version of his narrative on YouTube, curated by a faithful female fan called Maye Megg, who has a special interest in gay male relationships (complete episodes are also available on YouTube at the time of writing). Megg's sixty-six short videos, just a few of which have been blocked for copyright violation, reveal the intensity of the series' (and Yazpik's) investigation of a gay individual over some six months, an experience that is of course impossible to reproduce in the ninety minutes of a feature film.

Mauricio's character arc proves to be by no means predictable. Initially, at least, he seems highly unsympathetic. A strait-laced, financially focused student dressed always in a formal jacket and tie, he is set up as a foil to his hippie sculptress sister. It is significant in this context that Mauricio is not an artist, but an economist (as is his future lover). Homosexuality is placed from the start at the heart of the traditional family and of conservative values.

Mauricio is also provided with a fiancée who loudly voices her displeasure that after two years together they have not made love (the series is noticeably frank about sex and contraception). Here then the feminist argument for the activity of woman's desire (one that has no place in traditional telenovela) intersects awkwardly with the queer-positive argument for the discovery of repressed homosexual feelings. But Mauricio's initial straightness (in dress and conduct) is only partly a defense mechanism for his refusal of compulsory heterosexuality. It is also part of an attempt, as the actor said in his interview, to counteract stereotype and create a credible character.

Early on we hear of Mauricio's enthusiasm for a new visiting professor from Stanford at his university. It is in keeping with the painfully slow psychic process he is undergoing (typical also of the relaxed daily rhythms of telenovela), that it takes many weeks for us to meet Jim (born Jaime). He is a handsome bespectacled Mexican academic exile in California, who favors 1980s-style broad-shouldered jackets. His specialist research is on the effects of globalization on Latin American markets, surely a theme in which Argos itself takes an interest. So foreign is Jim in his home country that he is forgetting his Spanish and often slips into English (fortunately "homosexual," the word he uses in coming out to Mauricio, is the same in both languages). It is this ambiguously transnational figure from abroad (San Francisco, no less) who will become Mauricio's teacher in homosexuality as well as in economics. Mauricio even addresses him on occasion with only a touch of irony as "maestro" or "profe" ("teach"), even after they become a couple.

There are a number of key scenes here, accompanied by telling and varied televisual techniques that signal their importance. On Jim's first visit to the family home he makes an appraisal of the lush decor, singling out the cushions (he has bought similar ones in China!). The maid offers to bring the couple the typically Mexican *agua de jamaica* (chilled hibiscus tea). But when she comes back to the salon the camera lingers downstairs on the abandoned jug and glasses: Jim has asked to make a first visit to Mauricio's bedroom. Later Jim will discourse, also in the family home, on the nature of love, disagreeing with Mauricio who says that it has two faces, one internal and one external. As Jim makes his eloquent case for the true love that openly speaks its name, the camera cuts to a two shot of the siblings Mauricio and Diana, each equally seduced by the amorous pedagogue.

Finally comes the scene when Jim comes out ("I'm homosexual") to the still closeted Mauricio at the idyllic setting of a lake by the family's country house. The camera cuts back from the close-ups we might expect for such intimate dialogue to an extreme long shot where the two lonely figures are suddenly seen far off, silhouetted darkly against the water on a pier. This is the image of Mauricio's alienation and disorientation. And when Jim says he can never return to Mexico because of its discrimination against gays in the workplace, it is not hard to hear the series' progressive production team wishing that Mexican television too would open up to sexual diversity.

Mauricio is, as we have seen, characterized by sobriety. But fueled by a passion for Jim that he cannot at first acknowledge, he and his scenes later turn more melodramatic. In the swimming pool at the Country Club (a key location for middle-class Mexicans) the play of newly desiring looks between Mauricio and handsome fellow athletes is heavy-handed. When Mauricio first rejects Jim's advances he makes a scene, angrily exclaiming that they arouse in him only disgust. And when Jim subsequently threatens to return to Stanford and is about to leave for the airport, *La vida en el espejo*'s writers permit him, for once, an outright romantic cliché: Mauricio runs frantically through the street to Jim's home and embraces his departing lover outside the house.

Yazpik's most complex and famous scene is Mauricio's coming out to his father, now available on multiple YouTube sites. Its aesthetic is emblematic of significant moments in the series, whose careful visual style is far indeed from traditional telenovela. This is a lengthy sequence, lasting for some eleven minutes, made up of long takes, each of which are left to run for over a minute (a lifetime on regular television drama). And, strikingly, the sequence is wholly devoid of the music that is conventionally used to coerce attention and emotion from jaded or inattentive *telenoveleros*. Many of the shots are mirror shots, carefully framed alternately to reveal and conceal father and son from us and from each other.

It is a masterful, contained performance from the young Yazpik, very distant from the tearful histrionics of traditional television acting in Mexico. But the sequence's dialogue also serves as useful advice for viewers thinking of taking the same step as their beloved character. Mauricio's father, who is, as we have seen, an antipatriarch and no old-style macho, still asks the classic questions: "Why have you changed?" "When did you know?" "What did I do wrong?" (Mauricio's plausible answers are: "I haven't changed." "I always knew." "You didn't do anything.") By the end of the sequence the son seems to have inched toward reconciliation with the father from whom he was estranged. And in a world of generalized social simulacra, in which everyone is playing a part, homosexuality comes to stand here as a rare sign of existential authenticity: the weight of the mask he has worn until this moment, says Mauricio, is heavier even than the parental disapproval he so fears now.

Yet the mirror motif still troubles. Earlier in the sequence Mauricio had said that as a child, baffled by the inversion of the mirror image, he felt that his heart must be on the wrong side of his body. When human life is drawn into representation (in the mirror, on the television), it holds the promise, as telenovela knows so well, of both endless pleasure and endless pain as we attempt to fit image and reality together. We can now go on to see how another gay character is treated twelve years later in a second Argos series on the theme of the crisis of modern Mexican man.

The Dereliction of Masculinity: *El sexo débil* ("The Weaker Sex," Cadena 3, 2011)

The Camachos are a family with three straight adult brothers and a father (Álvaro, Dante, Julián, Bruno, and Agustín) who all consider themselves as being macho. On a single day, each of the Camacho's women leaves their respective partners. Doctor Álvaro's wife abandons him because he is jealous of her success at work. The girlfriend of Dante, a therapist, leaves him for a man she has met in Paris. Julián, a plastic surgeon, is abandoned by his fiancée because he's been constantly unfaithful to her. Their father Agustín is left by his wife on their thirtieth wedding anniversary when she realizes that he has neglected her throughout their marriage. The youngest Camacho brother, Bruno, is a gay social worker. Living as he does with his working-class lover, he is now the only one who is still in a long-term relationship. The abandonment of the male members of the family coincides with the arrival of Helena, a woman who has just left her wealthy fiancé at the altar rather than be trapped in a traditional marriage. As the new administrator at the medical clinic where the men are based, she will interact variably with the three sons and their father.

Like *La vida en el espejo*, which followed hard on the heels of *Mirada de mujer*, *El sexo débil* is a sort of male sequel to a female original, in this case Argos's previous hit for Cadena 3, *Las Aparicio*. There, three sisters and a mother are left without men after the loss of their male partners (one of them is lesbian and will marry her girlfriend at the climax). In *El sexo débil*, three brothers and a father will find themselves without women after they are abandoned by their wives or girlfriends (a further brother, Bruno, is gay). The teasing taglines for the two series are complementary, suggesting that female self-sufficiency requires a rethinking of an exhausted male identity. They are respectively: "A whole woman doesn't need another half" ("Una mujer entera no necesita media naranja"); and "If we don't stop being macho, we'll keep being the weaker sex" ("Si no dejamos de ser machos, seguiremos siendo el sexo débil").

It is *La Jornada* once more that presents a link between the two shows, not just in their shared director and writer, but in a recurring minor character played by Marco Treviño, who reappears here as he did in *Las Aparicio* as the ghost of a deceased husband (Cruz Bárcenas 2011). The article also suggests a hidden gay history of film and TV in Mexico. Thus, twenty-seven years earlier Treviño has starred in Jaime Humberto Hermosillo's *Doña Herlinda y su hijo* ("Doña Herlinda and Her Son," 1985), which he proudly claims as the first openly gay film in Mexico. Yet he complains that although Hermosillo's film ran for six months in London, it was barely seen in its home country. Treviño also defends the much-maligned medium of television (saying it is a great school for actors) and more particularly commends scrappy Cadena 3, which was later to win a franchise for a new national network under President Peña Nieto's broadcasting reform. The minority channel is, he says, putting pressure on the "mediocrity" of the two dominant networks (by now Azteca is as conservative as Televisa). The fact that Treviño also appears in the third text in this chapter, women's prison drama *Capadocia*, is further evidence for the role of actors in continuing LGBT-focused audiovisual content over the decades, a phenomenon no doubt noted by eager fans as well as scholars like myself.

This time it fell to *Milenio*, a centrist daily, to give the best account of Argos's new telenovela. Three years after it finished, the still mindful journalist offers a reiteration of the "macho imperatives" exploded by the series: that men have to be "strong as an oak tree," monosyllabic, careerist, neglectful of female desires, and invariably unfaithful to their unhappy wives. Also stressed (as already in *La vida en el espejo*) is Argos's formal distancing of its series from Televisa-style telenovela in its narrative structure (which appeals to "seasons" within the single series) and camera techniques that come close to those of art cinema.

But the main importance of a show for the journalist is the presentation of a loving stable relationship between two men (they will be called Bruno and Pedro). The characteristics of this gay couple are four: they are in their thirties; they run a CSO (civil society organization); they are committed to their own love story, however conflicted it may be; and they are, crucially, far from the cliché of the effeminate *joto* or drag queen. For the writer, this series that is intended for a relatively narrowly targeted audience points nonetheless to the possibilities of wider social change. It is striking, however, that he does not mention one clear innovation in this context: social class. While *El sexo débil*'s family are upper-middle-class professionals like those in *La vida en el espejo*, Bruno falls for Pedro, a working-class man whose roots lie far from his own reassuringly well-heeled milieu (Guadarrama Rico 2014).

Another change from *La vida en el espejo* is that twelve years later, promotional materials are much more polished and explicit. The synopsis calls explicit attention to the gay character. And the initial teaser video consisted only of a homoerotic image that is key also in print media: Raúl Méndez as psychiatrist Dante, the brother who is abandoned for a no doubt taller and blonder Scandinavian, is seen from above as he lies naked on a bed in fetal position.

Video promos for each character, which feature special footage not shown in the series itself, are also stylish, making use of elegant slow motion. The patriarch Agustín is seen throwing a glass of whisky into his gay son's face when the latter dares show open affection to his boyfriend at a family dinner. But the "oak-like" father is also wounded. He has, we are told, always been afraid of solitude, even in the course of thirty years of marriage. Gloomily, he balances a bare-breasted table dancer in his lap. Or again, the promo for plastic-surgeon-brother Julián (played by Mauricio

Ochmann, a major star of traditional telenovela with more Twitter followers even than Epigmenio Ibarra) consists of quick cuts of him making love to a succession of interchangeable sexy women. He even sends a text to the next girl before the last has left his bedroom. It is something of a shock to see Ochmann in such a role (one wonders how his self-named "Ochfans" dealt with the series), especially when the voiceover to the promo claims that, far from being a success in bed, he is "weak" because of his fear of commitment.

Such modern marketing no doubt helped the series' projection outside Mexico, which went beyond Televisa and Azteca's conventional alliances with US Spanish-language networks Univision and Telemundo and the subsidiary Azteca América (see Piñón 2011). When screened on NBC Universo, which bills itself as a "modern general entertainment cable channel for Latinos," it was called "daring, revealing" ("*El sexo débil*" 2015). Bilai Joa Silar, senior vice president, Programming and Production, called attention in the press release (like the representatives of Argos and Cadena 3 before her) to the quality and innovation of the series: "*El sexo débil* defies typical dramatic portrayals of Latino characters in Spanish television. It's a provocative series that challenges the traditional roles played by Latinos, and puts them in a modern light." Or again: "This fresh, edgy programming is representative of the quality entertainment NBC UNIVERSO is committed to offering our Hispanic viewers."

It is a kind of language rarely used by Mexican critics and scholars like Cueva and Orozco about their own television. But here, then, *El sexo débil*'s rare emphasis on antipatriarchal and antihomophobic themes (more explicit than in the case of the earlier *La vida en el espejo*) lent the series exportability outside its home country. US executives thus targeted *El sexo débil* at the quality demographics who prove resistant to the perceived limits of traditional telenovela (what Silar calls "Spanish television"). And in this reinscription, modern Mexican characters (especially gay ones) are flexible enough to become transmuted into "Latinos" or "Hispanics" for the US market.

What of the show itself? Early episodes, again as in *La vida en el espejo*, embed the gay plotline in familial and social contexts. And once more the queer story is somewhat anomalous (after all, the gay brother is the only one of the four who is not initially abandoned by his partner). The credit sequence signals the series' distance from Televisa-style telenovela. Rather than showing in traditional style the faces of the cast in an exercise of memory management for an inattentive audience, it offers a stream of impressionistic monochrome images, related only tenuously to the plot: a close-up of a bridal gown and posy, a pencil with broken lead, a child's sneakers in the rain.

Coinciding with the theme of subjective dissolution we saw in *La vida en el espejo*, the link between many of the images is water: water flooding over architectural plans, water pouring over a shirtless man in bed, water exploding in droplets on an outstretched male hand. Similarly, the music here is incongruous in the context of Mexican TV melodrama. We hear not a romantic ballad, as one would expect in telenovela, but rather an instrumental number with a funky guitar and bongos, signaling no doubt metropolitan sexiness and sophistication.

In the opening scene, sympathetic psychiatrist Dante (played by film star Raúl Méndez) waits in vain for his girlfriend at the airport, while Helena (Televisa veteran Itatí Cantoral) has hot sex with her fiancé in a baggage storage room. From the start, then, men's sexual and romantic experience is frustrated, women's active and pragmatic.

A series of quick cut sequences next lays out the lives of the father and heterosexual brothers for us, showing why their partners will abandon them. Patriarch Agustín ignores his wife ("the woman behind the great man") as she lays out his suit and shirt for him and shines his shoes. Gynecologist Álvaro reads the paper as his wife struggles to wrangle their children and get ready for work. Plastic surgeon Julián undresses and makes love to a new patient in his consulting room (she will prove in reality to be the steady girlfriend to whom he will offer an engagement ring that she accepts with evident ambivalence). Meanwhile, Helena, a medical administrator and specialist in male sexual problems, tries on the fairy tale wedding dress donated by the future mother-in-law who has arranged an "intimate" event, with just three hundred guests. Dramatically, Helena will flee from the future that has, she says, been "scripted" for her only at the ceremony itself.

More so even than *La vida en el espejo* twelve years before, *El sexo débil* takes as its starting point the breakdown of the sex/gender system embodied by the Cinderella narrative of the traditional telenovela. Hence, in a dramatic scene parallel to Helena's jilting of her perfect groom at the altar, the Camachos' mother walks out on her respectable husband at their thirty-fifth wedding anniversary party. The tone of the show may often be humorous, as when Julián addresses his penis, suddenly unresponsive after his fiancée ditches him: "Julito, we will rise again!"; or again when his brother Bruno describes himself jokingly as: "Just as gay as Ricky Martin . . . but better looking." But in line with the series' overall premise of male crisis, the five episodes shown over the first week of the season offer some uncompromising and disenchanting plotlines for both hetero- and homosexual viewers.

Thus, teasingly, when we first see Bruno he is posing as a client in a luxury brothel specializing in underage girls. Only later do we learn that he has come to that place in order to save one of their number. The theme of sex trafficking will be developed over a number of episodes, which involve several family members. First Bruno rescues the girl, María, spiriting her away from the short-stay hotel where he has arranged to meet her (it is typical of the series' sense of place that the dialogue uses the real name of the Hotel Cozumel, an actual motel in the reputedly dangerous *colonia* of Doctores).

Bruno then takes the girl to the community center run by him and his tough-looking boyfriend Pedro, who is a native of the grimy neighborhood in which their extensive workplace is located. The use of such authentic locations lends a powerful reality effect from a series that takes place far indeed from the overlit studios of Televisa. Slender, sexy, and very young, María is next lodged with psychiatrist Dante, who is troubled by his attraction to her. And in the final episode of the week, Dante and Helena go undercover at a glamorous party placed quite precisely in the upmarket neighborhood of Bosque de las Lomas. Their aim is to help María rescue four of her fellow girls from forced prostitution.

Another downbeat plot of the week concerns a terminal patient of Agustín, the father of the Camachos, who seeks reconciliation with the gay son he has previously beaten and become estranged from. There are multiple ironies here. Although Augustín encourages the man in this process, he does not tell him that he too has a gay son. And when he asks that son, Bruno, to intercede with the patient's son, Bruno sarcastically asks if he thinks all gays have some "secret code" for communicating with each other. But the series shows to its credit that even in the shadow of death, homophobia can still rule. Coaxed to meet up with his son and his lover, the patient, initially emollient, finally unleashes a torrent of abuse, shouting that the "disgusting things that queers do to each other" cannot be compared to the love between man and wife. He promptly drops dead. Bravely, then, *El sexo débil* refuses any easy resolution to deep-rooted prejudices. And we are left to wonder if Agustín, the patriarch of the Camacho family, who looks on at this drama, has learned something about himself and the son whose boyfriend he too refuses to accept.

It is not just homophobia but also class prejudice that makes Agustín hostile to his son's long-term partner. It is because Bruno fell in love with Pedro that the former abandoned a promising medical career, like that of his brothers, to engage in a dangerous brand of social work in a violent part of the city. By rejecting his medical mission, Bruno has (or so his father feels) betrayed his family. And the clinic shared by the other brothers (an ample house in upmarket Coyoacán, which was also used as the home of *Las Aparicio*'s female family) will serve the same role as the radio station in *La vida en el espejo*. The medical profession, like its media equivalent, offers a chance for the exploration of the conflict between the profit motive and the public good that is so dear to the hearts of Argos's Leftist team (CEO Epigmenio Ibarra gets an executive producer credit on *El sexo débil*).

As in the case of *La vida en el espejo*, a faithful fan or "shipper" (devotee of a fictional relationship), here known as Gaba2906, has once more collected the fragments of the gay couple's life in a lengthy YouTube playlist. This enables likeminded viewers more easily to plot the course of Bruno and Pedro's affair over more than one hundred episodes. What is most striking, however, and surely the result of the series being produced a decade after its predecessor, is that the problems of self-discovery and coming out to others, central to Mauricio and Jim's narrative, are here over before the series even begins.

We learn at the start that Bruno came out long ago to his family. And the unapologetic Pedro proclaims at one point that the couple are not "fags" ("maricones") but "proper queers" ("bien putos"). Countering stereotype, Pedro's working-class father had no problem with his son being gay. And it is established that the couple have lived happily together for years. Sharing coffee in bed together after a kiss (Bruno's furry chest is prominent on screen) the shaven-headed Pedro says that everything is perfect with them "as long the Camacho doesn't come out" in his partner. The only real problem remains Bruno's father's choice to tolerate but not accept his son's lover (his mother and most of his brothers have no problem with Pedro), a tension that will prove central to dramatic conflict in the series.

Yet it is vital to Argos's quality brand that characters, especially perhaps gay characters, should exhibit complexity and conflict within themselves. Thus, in the first episodes, Bruno, so cool and controlled when he rescues the girl from the trafficking mafia, is also hotheaded and foolish, jealously punching out a man who he sees talking to his boyfriend in the street. Typically shown in action sequences and formidable in a fight, Bruno also suffers from hidden frustrations, given to pounding a punching bag on the community center roof, where we see him sexily sweaty. If *La vida en el espejo*'s Mauricio is repressed, *El sexo débil*'s Bruno is all too physically expressive. He is a soft-spoken yet macho gay who, ironically enough, shares problems with his straight brothers. The philandering Julián, who is relatively supportive of Bruno, is also pathologically jealous of his female partners.

As the series develops, we see different aspects of Bruno's character. Often this is in the context of the classic tradition of "social message telenovela," which sought, as Guillermo Orozco suggested, to educate the viewer on contemporary issues. Thus, the gay couple start to discuss adopting a child, a natural extension of the nurturing process involved in their work helping trafficking victims, abused wives, and drug addicts. Bruno, a caring uncle, invites Álvaro's son who is experimenting with marijuana to observe the rehabilitation process that he offers to drug abusers in his urban Social Center, far from the lush suburbs where the teen lives. Taking pity on

Helena, who Agustín has deprived of work and with whose "discrimination" Bruno identifies, the gay couple offer to be her first patients at the clinic and submit to a prostate examination. Finally, they stand up to the drug-trafficking uncle of a young client who wishes to beat his addiction with them. The uncle taunts the pair in the street as *jotos* and says local kids should watch out that the couple don't try to turn them gay.

Beyond these social issues, Bruno is, as the series goes on, provided with a more detailed psychological backstory (we remember that we are not shown how he and Pedro first got together). We learn gradually that Bruno's first boyfriend died in his arms of a cocaine overdose (hence his exaggerated fearfulness and possessiveness over Pedro). And previously Bruno had gone out with a girl he felt he truly loved but did not desire. It is a secret he has never told Pedro. This more emotional register prepares us for an extended melodramatic plotline whose importance is marked by some rare camera techniques. When the homophobic drug trafficker knifes Pedro in the street, the camera spirals up in a crane shot to show his bleeding body spread-eagled on the pavement below. And in an unusual flashback or fantasy sequence Bruno seems to see Pedro playing soccer outside the center, although he and we know Pedro is now mortally sick in the hospital.

The hospital is of course a staple location of traditional telenovela à la Televisa. And *El sexo débil* turns the melodramatic screw yet tighter than conventional dramas, putting the father's relatively discreet homophobia to the test in this extreme circumstance. At first only Bruno's mother and one brother keep vigil with Bruno there, the other family members keeping their distance. But when Pedro goes into cardiac arrest, Bruno asks his father to operate, saying he wants to put the life of the man he loves into his hands. Finally, the father accepts. He even, hesitantly, embraces his weeping son after successfully performing the operation. A group shot then shows the three characters together: the lover lying with an oxygen mask in bed in the foreground with the weeping son and the moved father behind him. As Bruno's mother comments hopefully, surely this will mark a new kind of relationship between her ex-husband and their gay son? Even in this quality series addressed to a minority audience, melodrama can be made to serve the twin purposes of family reconciliation and social pedagogy.

El sexo débil is more reticent about lesbians than it is about gay men. In one lighthearted subplot, Julián's ex-fiancée successfully wins a bet that it is she and not he who can pick up a girl that both meet in a bar. In my final series from Argos, however, the lesbian theme is central to the plot and is treated with implications that are at once fascinating and disturbing.

Women behind Bars: *Capadocia* (HBO Latin America, 2008-12)

Capadocia tells the varied stories of a number of different women who have been sent to the (fictional) model prison of the title in Mexico City. Main characters among the inmates include La Bambi (a hardened criminal), La Colombiana (a former sex worker and drug dealer), and Lorena Guerra (a middle-class housewife who has accidentally killed her husband's lover). Within this private prison two members of the staff are also in constant conflict. Teresa Lagos, a married woman separated from her politician husband, is the new idealistic prison director, whose aim is to rehabilitate her charges. Federico Márquez is the (gay) representative of a private company that will secretly put the prisoners to work sewing lingerie with drugs concealed in the clothing. These political stories (public vs. private, rehabilitation vs. exploitation) are complemented by personal plotlines that include La Bambi's and Lorena's love rivalry over La Colombiana and Teresa's sharing of a young boyfriend with her teenage daughter.

The credits of *Capadocia* are abstract and enigmatic, showing female body parts and underwear, sewing machines, and somber images of an urban landscape lit by the moon. And the theme music is wordless, a simple acoustic guitar refrain leading into a haunting female choir. Both credits and theme tune serve to mark the difference once more with Televisa-style telenovela. And of course, *Capadocia* belongs to a very different genre: it is a weekly (not daily) series of just thirty-nine episodes spread over three seasons.

Shown over a period of years, but infrequently accessible, *Capadocia* served as "appointment viewing" for its elite audience and was by no means "another member of the family" like regular Mexican melodrama. (Most of the complete episodes are now generously shared by fans on YouTube.) Differences in distribution mirror those in content. *Capadocia* benefited from a larger budget and greater explicitness in sex and violence than is possible on generalist television: the opening sequence is of inmates naked in the communal prison shower. And the technique aspires to that of the art movie. The series was shot with a relatively leisurely schedule on film, not digital video, and in a full-size prison set built for the purpose in Mexico City. Promotion was also professional. Expert publicity shots show the cast tied up with red rope, an arresting image of their confinement which does not appear in the show itself.

What of the production process? This was the first HBO show to be made in Mexico, something of a landmark for all concerned. But perhaps this was a problem for anti-American Leftists at Argos? Yet, as with Argos's previous shows, *La Jornada* supported the producers' launch of their new title ("*Capadocia*, historia" 2008). Appealing to cultural nationalism, it identified the series as a "Mexican production" and stresses the show's "documentary realism" when it comes to depicting contemporary Mexico. Epigmenio Ibarra, whose contribution to the premise was the idea of a privatized prison (in fact unknown in Mexico), proclaims that the series is "not telenovela, but cinema," thus attempting to secure its special status. And the series' Leftist critique of corruption in modern Mexico is in line with *La vida en el espejo*'s skeptical discussion of mass media and *El sexo débil*'s reappraisal of for-profit medicine.

Yet of course *Capadocia*'s audience was more restricted than its predecessors. While Argos's previous shows were broadcast free to air (although Cadena 3 has a smaller audience than the big networks), this one was for a pay subscription channel. Its prestige derived from a canon of quality TV series in the United States, well known also to educated audiences in Mexico. Nonetheless, *Capadocia* was targeted not only at its own country of production but also at a continental audience. It was shown on sister channel HBO Latino in the United States (where it won a GLAAD award for LGTB depiction) alongside a broad range of Spanish-language content, including movies from distant Spain. But it was also directed, as the name of its affiliate suggests, to a whole continent. Previous series from HBO Latin America, also focusing on the crime genre and boasting the requisite doses of explicit sex and violence, were shot in Brazil and Argentina.

I would suggest, however, that queer TV auteurship should in this case be reassigned from HBO to Argos. We have seen that the Mexican production company addressed similar themes to those of *Capadocia* in earlier series, including a pioneering focus on LGBT characters. *Capadocia*'s versatile screenwriters included Leticia López Margalli, who had contributed long before to *La vida en el espejo*. And, as mentioned earlier, Marco Treviño, the ghostly Aparicio husband in *El sexo débil*, reappears in *Capadocia* playing a politician who is separated from his wife Teresa. Currently the mayor of Mexico City, he is standing for election as president of the republic, no less.

HBO's trademark dirty realism and cinematic ambitions are evident in the opening episode. They are seen most especially in an expertly choreographed large-scale prison riot in the old prison, which precedes the new high-tech penitentiary to be called Capadocia, after the supposed birthplace of the Amazons. Yet the social critique in the premise that is set out in the synopsis chimes once more with the corporate mentality of Argos, as much as with HBO. The prison director Teresa, focused on reform and rehabilitation, is not so different from the idealist social workers of *El sexo débil*, albeit working on a much larger scale and in a more prominent position. Much of the tension in the series' central plot arises from her conflict with colleague Federico, who conspires with corrupt politicians to put the inmates to work stitching illicit drugs into lingerie.

Federico is played by Juan Manuel Bernal, the distinguished actor from *Cuatro lunas* who, as mentioned in the previous chapter, was outed by gossip magazine *TV-Notas*. The fact that his character here is gay represents something of a step forward (or perhaps back?) for Argos's continuing project of queer TV: while *La vida en el espejo*'s Mauricio is repressed but moral and *El sexo débil*'s Bruno macho but idealist, *Capadocia*'s Federico is an out and out scheming villain, with no redeeming features whatsoever.

Subplots extend the range of LGBT characters in the series, playing a useful role in Argos's social pedagogy once more. Teresa struggles to get a lesbian couple the conjugal rights of which they are denied or to have a transgender woman transferred to Capadocia from the men's prison to which she has been confined. But the main queer theme is lesbianism within the prison itself. Voluptuous La Colombiana (a veteran of drug trafficking) is not just loaned out to heterosexual prostitution by the prison authorities (she briefly escapes into the city), but is also handed over as booty to the prison's current female kingpin. She passes successively from La Regina, a boss dethroned by the opening riot in the first episode, to La Bambi, an enamored psychopath who manages to follow her to the new prison, and Lorena, initially a terrified housewife (our identification figure) who is confined to Capadocia when she accidentally kills the lover of her erring husband.

It is with Lorena, played by Ana de la Reguera, a TV veteran in the United States as well as Mexico and a major movie star in her home country, that HBO-style realism collides most forcefully with Mexican-style telenovela. In the opening sequences of the first episode, Lorena's comfortable middle-class kitchen, where she makes a breakfast of pancakes for her husband and kids, is pointedly contrasted with the grungy hellhole of a prison, where the food is less than appetizing. And her later discovery of her erring husband in bed with her best friend is pure melodrama. As she is led away by the police, her little children ask: "Where are you going, mommy?" It is a moment that would not be out of place on Televisa or Azteca's prime-time schedules. Yet, as in Argos's previous series, the sex-gender system is shown here to be in complete disarray from the start. When Teresa's marriage is shattered also by her adulterous husband, she (like Isabel in *La vida en el espejo*) takes on a much younger man as a toy boy. Unsurprisingly, this does not turn out well for either of them.

Such soapy plotlines are masked by a technique that aspires to art movie status (we have seen this aspiration also in Argos's series for free to air TV). *Capadocia* abounds in expert long takes or sequence shots, striking angles (as when the camera looks down on the battered Bambi, killed by Lorena from on high), and disorientating fantasy sequences (Lorena imagines she is making love to her husband while in fact she is being caressed by the sinister old woman with whom she shares a cell). The central lesbian theme might be read in itself as part of the aesthetic and thematic sophistication that the series uses to mark its difference from ordinary Mexican TV. After all, as we have seen, lesbianism is not as common as male homosexuality in Argos's earlier shows. The big exception (untreated here) is *Las Aparicio*, which boasted the first lesbian wedding on Mexican TV and was available, like *El sexo débil*, without payment on Cadena 3.

To its credit *Capadocia* avoids the alibi that its imprisoned women make love to each other only because they have no access to men. On a tearful visit to the cemetery, La Colombiana reveals that scary Bambi was the only partner, male or female, who ever loved her. Lorena's own slow assumption of lesbianism is shown to be an essential part of her character arc in the first season, as she moves from submissive housewife to hardened prison top dog. The limited, special circumstances of incarceration thus give way as the series develops to an expansive national allegory with very wide implications. Just as *La vida en el espejo* critiqued commercial mass media and *El sexo débil* investigated the economics of medicine, so (and much more explicitly) *Capadocia* explores the privatized prison as an image of corruption and claustrophobia. It is a vision felt by many, especially on the Left, to be characteristic of contemporary Mexico.

This sense of generalized confinement is caught in the shooting style of a three-minute sequence thirty minutes into the last episode of the first season. First the camera pans slowly over an ominous blood red sunset. Black electricity pylons are visible on the horizon. We cut to the sterile white bathroom at the home that progressive prison governor Teresa (who is, as mentioned earlier, separated from her politician husband) shares with her two daughters. The elder teenage daughter steps out of the shower, shielding her body with a towel (the inmates are given no such cover in their shower sequences). The camera pans left as we see the daughter's hand demisting the bathroom mirror. What she sees, however, is not herself but an image of her boyfriend kissing her mother: the fantasized Teresa even looks out from the mirror, smiling with complicity at the daughter. The latter has just learned that both she and her mother have slept with the same man.

Next we see the daughter, fully dressed in black, stalking down the corridor at home. From behind, her mother and younger sister emerge, clad in pink and white, holding a birthday cake and singing "Las mañanitas," the Mexican equivalent of "Happy Birthday." In standard shot reverse shot we watch and listen as the mother encourages the daughter to stay "with us" (the Spanish pronoun is explicitly feminine "nosotras"), while the daughter claims an urgent engagement with a friend.

Suddenly we cut to the large dining room in the prison. While Teresa's daughter rejects her lovingly offered cake, here two blue-clad inmates, seated at a small round table, do not eat the food in front of them, complaining it is stale. And they ask the older lesbian who joins them when she thinks they might be released. She tells them definitively: "We've got to get used to the fact that we won't be on the street for many years." What is striking about the scene, however, is that unlike the smooth rhythm of the shot reverse shot of the previous scene, here the camera moves continuously around the three prisoners in a dizzying circular tracking shot.

In its very self-conscious technique (the blurring of reality and fantasy in the bathroom, the vertiginous camerawork in the dining room) *Capadocia*, unlike my previous series, calls attention to its aspirations to artistic distinction. Yet TV form is not simply decorative here but rather expressive of deeper concerns. Although the shooting style may be different, elements of mise-en-scène link these two locations that are juxtaposed by the editing. The three female figures in Teresa's comfortable home mirror the threesome shown next in the brutal prison setting. And a further graphic match links the two scenes: a single blue wall in the home foreshadows the blue uniforms in the prison. With or without bars, the same-sex female households of the series, apparently so different, are equally claustrophobic.

Yet, as ever in textured quality television, such a process is complex and ambivalent. And in the final sequence of the same last episode, Teresa's teenage daughter will be confined to Capadocia after a botched birthday raid on a jewelry shop in the Zócalo (hence her black attire in the previous sequence). Now it is the young girl who is joined in bed by the old lady inmate. This is a disturbing image of the predatory lesbian that deserves to be read, however, within the pervasive cruelty and degradation of the fictional world depicted in the series, which spares none of its many varied characters. Homosexuality served in *La vida en el espejo* and *El sexo débil* as a rare example of authenticity and morality among entrenched hypocrisy and criminality. In *Capadocia*, conversely, queerness proves to be no escape from the confines of a depravity that is presented as the universal condition of modern Mexico.

Intimate Strangers

The second season of *Capadocia*, like the first episode of *El sexo débil*, begins with a runaway bride. But here the sequence is cinematic in style and grave in subject matter. A young woman is pursued by a helicopter through a forest in her lacy gown and veil. She will prove to be the new wife of a drug trafficker, fleeing the slaughter carried out by a rival narco at their wedding. Soon, still in her princess-style dress, she will be banged up in prison. Her new comrades warn her about the old lady who seems so friendly: she's a dyke (*tortillera*).

Clearly there are family resemblances between the three series I have treated by the same production company, in spite of transparent differences in format, production processes, and distribution. We would seem to be a long way from the blatant censorship and stereotyping of *Vecinos*, Televisa's adaptation of the Spanish comedy. But none of the three series studied here were broadcast by the still hegemonic broadcaster. And the *joto* stereotypes of *La hora pico* were still trotted out at the 2015 *TVyNovelas* awards, where the queer waiter worked the crowd to the delight or embarrassment of the assembled celebrities. The magazine itself has moved on. When it carried a story on big-breasted singer Sabrina Sabrok, a regular on Mexican TV, including *La hora pico* and LGBT talk show *Guau*, it was to shame the "gay ambassador" for allegedly rejecting her lesbian daughter ("Sabrina le hace el feo" 2015).

The change in attitudes that we saw in the Mexican film establishment in previous chapters is perhaps permeating television also. Such a change would be likely to have a substantial social effect, as the small screen plays a much more central and durable part in national emotional life than the big screen, as shown by middle-aged viewers' still-fond memories of Yazpik's troubled Mauricio. Perhaps if there were more LGBT characters on television, there would be less pressure for them to be "decent" (code for "straight-acting") like Mauricio, and even a possibility of reevaluating effeminacy. Meanwhile, Juan Manuel Bernal's Federico in *Capadocia*, the businessman who mercilessly exploits the inmates and is allied with both corrupt politicians and narcos, is a bracingly different kind of gay character. He was no doubt welcomed by HBO's elite audience, accustomed as they are to protagonists who are complex, difficult men.

Although its career in quality TV is to be celebrated, Argos, which made the series treated here, is just one producer among many on the Mexican TV scene. Yet it, like Televisa, has its own acting school (called CasAzul), an initiative that may perhaps help to damp down the declarative performance style so common in telenovela. Among its teachers are Aida López, who plays a steely prison guard, the ally of gay Federico, on *Capadocia*.

It may be true for the moment that Televisa and Azteca still have a stranglehold on the free to air TV that entertains all but the upper classes each night in prime time. And Televisa's star system, bolstered by print magazines and awards that are rigorously restricted to its homegrown performers, remains formidable. But producers and channels other than Argos have also broken through this stalemate. Educational Canal 11 has broadcast a number of quality series including one, *XY* (Nao Films, 2009–12), whose take on troubled masculinity is very similar to that of *El sexo débil*. *XY* features a moving and convincing portrait of a gay male couple with few rivals on Mexican television, much less film.

Of course, diversity cannot be reduced to sexual preference. My three titles fail to revise existing television biases in region and race. Neglecting the vast territory of a nation that stretches from the Pacific to the Gulf and from Texas to Guatemala, all are set and shot in Mexico City. And only *Capadocia* features darker-skinned or mestizo-featured actors, and even they are playing only working-class roles. Yet, I have suggested that the old televisual regime is starting to crumble. We saw in the introduction to this chapter that the audience for telenovela, already unattractive to advertisers, is in decline. And recently Televisa announced that after the final episode of retrograde remake *Simplemente María* (which will no doubt climax with the whitest of weddings) it will schedule no more telenovelas at four in the afternoon ("¡Se acabaron!" 2016).

As mentioned earlier also, with the government reform of broadcasting, a new digital network, somewhat delayed, will open, run by Cadena 3, the home of the self-consciously daring *El sexo débil* and *Las Aparicio*. The motto of this free to air channel is "More open than ever" ("Más abierta que nunca"). It is to be hoped that, as with the initial appearance of Azteca in the 1990s, competition will lead to diversity and innovation in fiction. It will make little difference to viewers if that creative change is motivated by the commercial aim of reconnecting with the newer quality audiences who have turned their backs on local TV. The conservative television apparatus thus joins the macho sex/gender system as one of those outdated or "residual" Mexican traditions that is ripe for reform.

This chapter has turned out to be the longest in the book. Its length no doubt corresponds to the extensiveness of the object it treats (telenovelas have over one hundred episodes), but also to that object's complexity. It is the nature of quality television drama, in spite of its necessary recourse to repetition and familiarity, constantly to work through social issues while employing multiple viewpoints and avoiding definitive conclusions. It is for this reason that if the telenovela is, as Rubén Jara claimed, just another member of the family (as Mexican as green sauce), television itself remains an intimate stranger. In this it is like those queer kids who gather still around the set with their unsuspecting parents and whose stories may also prove to be unpredictable and unfinished.

It is striking that in the shows I treat here (unlike in the great majority of the feature films in the earlier chapters), lesbian and gay characters are admitted into the family home, internalizing in the plot the TV medium's domestic mode of consumption. On *La vida en el espejo*, Mauricio's father is angry because he finds out that the professor Jim, an honored guest in his house, is not just gay but the lover of his student son. On *El sexo débil* the working-class boyfriend Pedro is made unwelcome at the fancy but disastrous anniversary party thrown by Bruno's parents. And on *Capadocia* Lorena is asked by the ghost of the Bambi she has killed (the previous lover of La Colombiana) how her children will react back home when they find out she became a "dyke" in prison. The integration of LGBT people into Mexican families and Mexican television will no doubt remain awkward, as it is elsewhere. But it is likely to continue to give rise to compelling, complex drama, whenever it is given the chance.

CONCLUSION
Two Films, Two Futures

It's a chilly evening in New York in March 2016 and I'm at the annual New Directors/New Films Festival at Lincoln Center. I'm here to see a Mexican queer fiction feature, *Te prometo anarquía* ("I Promise You Anarchy") from Julio Hernández Cordón (not to be confused with the established auteur Julián Hernández of chapter 2). Solitary gay cinephiles like myself line up meekly and lengthily for entrance as the previous screening is running overtime. Overwhelmingly male and often elderly, they are finally greeted with an announcement that patrons with canes or walking sticks will have priority access. This melancholy spectacle takes the hallowed ritual of cinephilia (studied in chapter 4) into new, unpleasurable territory.

Ironically this is a youth picture, whose unusual premise is of two male lovers and skateboarders, homeless in Mexico City, who sell their own (and others') blood to earn their precarious living. At its Mexican premiere in Morelia in 2015, where it won the press prize, Carlos Bonfil, who is, as we have seen, an acknowledged expert in the field, praised *Te prometo anarquía*'s "fresh" narrative and "vernacular" dialogue. He acclaimed it as an "urban fresco where social marginality and sexual diversity intersect" (Bonfil 2015b). Widely shown on the international festival circuit, the film also won a special mention in San Sebastián, an event that prides itself on serving as a link between Europe and the Americas. Appropriately enough, the film is a German-Mexican coproduction.

Te prometo anarquía does indeed seems fresh, avoiding the punishingly long takes and glacial pace of many Mexican festival films before it, such as Matías Meyer's *Yo*, which won the main prize at Morelia, also in 2015. There are frequent dynamic shots of the young couple skating through varied and iconic urban locations, from the grand Monument to the Revolution, recently renovated, to the crowded and chaotic Merced market. We also see them kissing through a lit window in the Hotel Cozumel, the place of assignation where the gay Camacho brother met the teenage prostitute in *El sexo débil* (see chapter 5). There is some audience-pleasing frontal nudity (even a naked skating scene) and a well-curated soundtrack of Mexican and foreign indie rock and rap.

Yet *Yo te prometo anarquía* still falls fair and square within the festival film genre intended primarily for elite foreign audiences at venues such as New York's Lincoln

Center (indeed it has not been theatrically released in its home country). A late plot twist has the couple selling blood to narcos, who then kidnap fifty of the human "milk cows" the gay boys have herded together (no such event has actually happened in Mexico City). And, inexplicably, the pair, hitherto harmless, suddenly beat to a bloody pulp the contact who had helped them set up the deal with the drug traffickers. The fact that this sequence takes place in the revered studios of Churubusco (where, as we saw in chapter 3, child star Pinolito took his bow and his alter ego Coral took a stroll with her aged mother) might be taken as a sarcastic wink to Mexican film tradition. But the film clearly places itself in the line pioneered by Cannes's favorites Carlos Reygadas and Amat Escalante: art movies seasoned with sex and ultraviolence for foreign cinephiles.

The fact that the sex (unlike in Reygadas and Escalante) is gay now seems indifferent. In spite of the prominent love scenes, lit by a glowing red light, the couple are not provided with the romantic halo or erotic heat they would have had in a Julián Hernández film. Nor are they furnished with the kind of psychological backstory carefully contrived for its queer characters by Argos's telenovelas. Sullenly alienated throughout (although one seems to come from a wealthy background), the pair would also prove unattractive to the mainstream audience I discussed in chapter 4.

Two weeks earlier I had attended in Mexico City a commercial screening of another queer-themed Mexican fiction feature, *Las Aparicio*. This new release was directed by Moisés Ortiz Urquidi, who also helmed episodes of Argos's telenovela by the same name, which, as we saw in chapter 5, served as a kind of female prequel to *El sexo débil*. The film kept the same feminist tagline as the series: "A whole woman doesn't need another half." And its poster, most unusual in a still-male-dominated film industry, showed only the five female actors, who play the cigar-smoking matriarch, her three daughters, and the wife of one of them, now promoted to principal cast member. This last performer (returning with her original lover from the series) is Eréndira Ibarra, the daughter of Argos CEO Epigmenio, who takes a producer credit for his company's rare venture into feature film.

Although TV fans like myself were no doubt disappointed by some changes in the casting (the central sister was replaced by another Argos veteran, Ana de la Reguera from the prison drama *Capadocia*), the film opened wide throughout the Mexican territory on a thousand screens. When I saw it at Cinépolis's multiplex on Bucareli Avenue (a central if still slightly slummy location), the audience was overwhelmingly female and the frequent lesbian love scenes provoked no audible comment from them or the rare male viewers. Unlike *Te prometo anarquía* (and indeed its original TV series, which trafficked in metropolitan sophistication), the film of *Las Aparicio* is set in a rural location, the ample mansion in tropical Veracruz that is the ancestral home of the extended family. Main plotlines include the lesbian couple's search for a sperm donor (the hunky gardener will do nicely) and the resolution of the curse hanging over the Aparicio women, whose male partners always die. It turns out that the sisters' great-grandmother, an early feminist heroine, shot dead the abusive husband she had been forced to marry and walled his body up in the family home.

The film of *Las Aparicio* punches up the lesbian theme that was already prominent in the series. And it is significant that the two actors playing the couple posed in publicity materials and at the premiere as if they were indeed lovers. While the sex scenes in the film itself read as soft-core porn, they are part of a strategy of visual and narrative pleasure that (as in my mainstream movies of chapter 4) place a queer look at the center of commercial cinema. And even in rural Veracruz, like in sophisticated Mexico City, now no one turns a hair at open lesbian displays of affection, at least within this stylish cinematic version of sexual diversity.

I would suggest, then, that *Te prometo anarquía* and *Las Aparicio*, shown in the same year, suggest two futures for Mexican queer film. The first takes foreign festivals as it main mode of exhibition and fulfills the aesthetic and thematic criteria demanded for and by such minority audiences. The fact that the action of the film is male-dominated (one of the gay boys has a girlfriend neglected by him and the plot) may be no accident, given the requisite presence of narco violence to a genre favored, abroad at least, by mature men. The second, *Las Aparicio*, seeks to seduce a younger, local, and female audience with a mix of glossy eroticism and queer politics. Indeed, one of the actresses stressed her film's "social contribution" to Mexico (Castañeda 2016). And while *Te prometo anarquía* follows the template of festival film, *Las Aparicio* appeals to a televisual aesthetic (now, of course, as polished as mainstream cinema) and a mass audience nurtured also by the dominant TV medium.

The gay subtext of much traditional Mexican television is not hard to discern. After still photographer Stefan Ruiz spent years chronicling Televisa's studios, he placed a shirtless hunk on the cover of the resulting volume, *Factory of Dreams* (2012). The unknown actor lolls on a bed in front of a transparently fake backdrop of city skyscrapers at night. I have argued in this book, however, that in spite of such pervasive escapism, it is television, which invites such figures as Ruiz's hunk into family households, that has the greatest potential for achieving a "social contribution" to a modern Mexico that can still be dangerously conservative.

Institutional changes in early 2016 may provoke changes even in the heritage networks. Televisa launched its streaming service, the oddly named and lower cased blim, by highlighting a quality period series some distance from traditional telenovela: *El hotel de los secretos* ("The Hotel of Secrets") is based on *Gran Hotel* (Antena 3/Bambú, 2011–13), a glossy Spanish original. Azteca, for its part, appointed a new young director general, who promised to shake up the perennial runner up's currently stodgy content, even if he was the son of the previous CEO (González 2016). One commentator wrote that there was now a good opportunity for the channel to return to the glory days of its collaborations with Argos and Epigmenio Ibarra (Moscatel 2016), which, as we saw in chapter 5, spawned the pioneering *La vida en el espejo* with its groundbreaking gay character.

If, as I have suggested, TV is to be reevaluated in a rapidly changing audiovisual context, the cultural distinction still traditionally ascribed to auteur film should also be reexamined and perhaps called into question. It is significant that the shorts by Hernández and Fiesco currently available on Amazon Instant Video are given a title (*Mexican Men*) and a promotional blurb that openly suggest the once-hidden promise of an erotic link between art cinema and porn, which has also been a theme of this book.

Future research for an earlier period than the one addressed in this book might reevaluate the sexy film comedies of the 1960s and 1970s, in which effeminate or *joto* characters are so prevalent. One well-known example would be *Modisto de señoras* ("Ladies Dress Designer") of 1969, which I saw on afternoons in both 2015 and 2016 on Televisa's XEQ Gala heritage channel (it is also available at full length on YouTube). Here, in a transparent disavowal, suave star Mauricio Garcés pretends to be gay in order better to seduce his buxom female clients, who are generally clothed only in negligees when he arrives for his house call. Incensed by this pretense, a trio of campy, catty professional rivals try to bring him down. Finally, however, Garcés's character (who goes by the fancifully French name of D'Maurice) joins up with his nemeses after being forced into an embarrassing but reassuring heterosexual coming out. But in the last sequence when all four are measuring up an impeccably macho muscular male for a suit, even he turns out to be queer. The conservative comedy thus brings with it an unexpected and radical final moral: gays are everywhere in modernizing Mexico and, unlike in the case of D'Maurice, who is in any case faking it, they cannot always be recognized.

Once more the television medium may be ahead of film here. The real-life singer Juan Gabriel is a rival to José José (Julián Hernández's favorite) as the romantic crooner of the 1980s best loved by housewives. When it was announced that Disney Media Distribution Latin America was making a series based on the life of the star affectionately known as "Juanga," the main news was that the thirteen episodes would "tackle the diversity of love" (Salgado 2016). Producers and actors alike stress their "pride" that now the true story of this national treasure can be told, unlike in a far off, benighted time when "sexual preferences were prohibited." And, remarkably, they have the full approval of the veteran, hitherto closeted, singer for this "faithful portrait."

As I argued in my introduction, then, there is evidence, both social and cinematic, for a new Mexico when it comes to queer visibility. Yet it is also clear that in an age of media convergence our object of study should be expanded beyond feature film and especially beyond the much-studied art cinema that is so little distributed in its home market to embrace the audiovisual field in all its rich variety. Only then can we start to piece together a picture of a queer Mexico that might tell us something not only about film and television but also about the vast and diverse nation itself.

APPENDIX
Interview with Julián Hernández at the Morelia International Film Festival, 2014

Julián Hernández (whose works I examine in chapter 2 of this book) is that rare thing, an openly gay Latin American auteur. After graduating from film school in Mexico City and while working always in collaboration with his partner-producer Roberto Fiesco (now himself a prize-winning director), he has made four features and over a dozen shorts over the course of a decade, all of which reveal an impressive commitment to both queer content and avant-garde filmic practice.

His first feature, *Mil nubes de paz cercan el cielo, amor, jamás acabarás de ser amor* ("A Thousand Clouds of Peace Encircle the Sky, Love, Your Being Love Will Never End," 2003) was nominated for multiple Ariels (Mexican Oscars) and was the winner of the LGBT Teddy award at Berlin. It charted in moody black and white the hopeless story of a part-time Mexico City hustler abandoned by his older boyfriend, all to the musical accompaniment of Spanish songstress and camp icon Sara Montiel. The title would subsequently be used for the name of the production company Hernández founded with Fiesco, Mil Nubes Cine. The company has since also produced films by both distinguished veterans such as Arturo Ripstein (*Las razones del corazón* ["Reasons of the Heart," 2011]) and novice directors such as Iria Gómez Concheiro (*Asalto al cine* ["The Cinema Holdup," 2011]), not to mention Fiesco's first documentary feature *Quebranto* ("Disrupted," 2013), treated in chapter 3 of this book.

Hernández's second feature, *El cielo dividido* ("Broken Sky," 2006) (also accepted by Berlin), staged a more romantic and intermittently sunny love triangle between three improbably cute university students, all shot in pretty pastel colors. *Rabioso sol, rabioso cielo* ("Enraged Sun, Enraged Sky," 2009), winner of the Teddy award once more, Hernández's third feature and over three hours long, marked a radical change. Its first half offers an uncompromisingly gritty vision of contemporary gay Mexico City, including an impressively dilapidated picture palace that is used as a cruising site by the film's erotically driven males. Its second half relocates the same characters (a gay love triangle once more) to the mythical setting of the spectacular canyons and pre-Hispanic ruins of Cadereyta in the province of Querétaro.

Finally, the tripartite *Yo soy la felicidad de este mundo* ("I Am Happiness on Earth," 2014), his most recent feature, is Hernández's most explicit commentary on both gay filmmaking and gay sex. The main character is an art movie director who has affairs with a dancer and a sex worker in the first and third sections, respectively, of the feature, while the middle section, a film within a film, shows us what appears to be a lengthy sequence of the character's work in which modern dance moves mingle with footage of real-life masturbation. *Yo soy la felicidad de este mundo* was released theatrically in the United States in August 2014 and was subsequently made available on Amazon Instant Video.

Julián Hernández kindly sat for this interview on October 22, 2014, at the twelfth edition of the Morelia International Film Festival, where *Yo soy la felicidad de este mundo* received its Mexican premiere and was selected for the best fiction feature competition.

Q: The title of *Mil nubes de paz cercan el cielo, amor, jamás acabarás de ser amor* cites a poem by Pasolini, and the dialogue of *Yo soy la felicidad de este mundo* refers to Fassbinder. A poster of the latter's *Veronika Voss* (1982) also hangs on the main character's wall. What is your relationship with gay European art movies? And were you aware of the New Queer Cinema in the United States that appeared in the decade before you made your first feature?

A: I've always been a cinephile, but before I went to CUEC [Centro Universitario de Estudios Cinematográficos, one of the two prestigious state film schools] I was more familiar with earlier Mexican cinema. And when I first got to know Pasolini, it was through his novels rather than his cinema. I was most impressed by German film: through Alexander Kluge I got into Fassbinder's films, starting with *Effi Briest* (1974). It's true that Pasolini and Fassbinder are fundamental for me. But what attracted me to the former was the literary aspect, while for the latter it was the use of the image. It was via Fassbinder too that I made the connection with melodrama, which I traced back to its use by [classic Mexican director] Emilio "El Indio" Fernández. Werner Schroeter was also initially a great influence on me for his operatic sense of cinematic structure in films like *Der Rosenkönig* ("The Rose King," 1986). I wasn't aware of the New Queer Cinema in the US until my first feature, *Mil nubes*, was shown in Berlin, where I acquired the same distributor as Gus Van Sant.

Q: In spite of these cosmopolitan references, Mexican locations (especially in Mexico City) are very prominent in your films. I'm thinking of the extraordinary vision of the chaotic capital and of mythic rural Querétaro in *Rabioso sol, rabioso cielo* or of the architecturally striking UNAM [National Autonomous University] campus in *El cielo dividido*. How do you see the role of this kind of authentic location in your films?

A: When I began my career, it was very common in Mexican cinema to shoot in the studio, and real locations were rarely used. For me it's vital to use real places. Indeed, they are the defining context of my films and a vital part of the creative collaboration with the actors. For example, in my second feature the campus of the UNAM served as a place of liberty and tolerance for my student characters. Although the film was criticized for not representing "reality," my aim was to depict a utopian space in which there was no homophobia. In my third feature, *Rabioso sol, rabioso cielo*, the characters are analogous to natural forces. It thus seemed natural to place them in the ancient locations of pre-Hispanic ruins of Toluquilla, although it was of course not easy to shoot there. In spite of this quintessentially Mexican setting, the visual reference here was in general to the figures in a landscape in Renaissance painting and more particularly to the role of the setting in the frescos of Masaccio.

Q: Beginning with *Mil nubes de paz* you have insisted on the use of popular song in your films, from Sara Montiel to José José (the so-called "Prince" of Mexican popular song). And you have even made a documentary feature with Roberto Fiesco called *La transformación del cine en música* ("The Transformation of Film into Music," 2009). Often there's a kind of disjunction between the romantic music you choose for us to hear and the desolate vision of gay romance that you have us watch. Tell us about your approach to music as a resource in your cinema.

A: I myself studied to be an opera singer, but I also love the masterful José José, whose music features regularly in my films. In my cinema, the songs are never decorative but serve to advance the narrative. While I tend to use rather simple plots, I employ music to create emotional atmospheres and to enable the audience to understand through the senses rather than the intellect. Moreover, following the role of music in melodrama, the lyrics often serve to voice the words that the characters themselves should speak but cannot. For example, in *Yo soy la felicidad de este mundo* the two lovers are closest not when they have sex but when they sit on the bed without touching, singing along to a romantic song together. This use of music is reminiscent of Fassbinder once more, but also harks back to the musical numbers in [classic Mexican directors of the Golden Age] Roberto Gavaldón and Julio Bracho. I'm not averse either to using recent musical sources that are considered to be of low status. In my current fourth feature,

I use music by Mexican pop singer Carla Morrison, who is not always thought of as a quality artist.

Q: In *Yo soy la felicidad de este mundo* a new theme appears in your cinema: dance. I wonder if, beyond the simple movement of the body in space and time, which is itself highly cinematic, you have some motivation for this new interest?

A: Back in 2000 I made a short just one minute long called *Rubato lamentoso*, which consisted of two figures dancing in a desert. Since then I've studied the code of dance, the code of film, and the nexus between the two. My new feature *Yo soy la felicidad de este mundo* opens with a dance sequence by veteran performer and choreographer Gloria Contreras, whose dance workshop I followed at the National Autonomous University. One of the three main actors in *Yo soy la felicidad de este mundo* is Alan Ramírez, who is a professional dancer and member of Mexico's National Dance Company. By including media such as music and dance in my cinema I'm suggesting that film can be like opera for Wagner, a total work of art.

Q: In all your films, there are attractive gay sex scenes that are relatively inexplicit. In your short *Bramadero* (2007), however, whose title translates as a "rutting ground" for wild animals in heat, there are sequences of real fellatio and anal penetration. How do you conceive the integration of real, not faked, sex into filmic narrative? Does it imply any special challenge or make any special contribution to the film?

A: I put into my films what I would like to see as a spectator. I really don't like tricks on the audience, as in *L'Inconnu du lac* ("Stranger by the Lake," Alain Guiraudie, 2013) or *Nymphomaniac* (Lars von Trier, 2013) where the directors used body doubles or digital effects for explicit scenes. I think of such sequences in my films as deriving from a necessity for the director rather than as posing a challenge to the actor. And, as I mentioned earlier, one of the actors in the new film is a dancer and thus trained to use his body. I think if you trick the audience with fake sex this produces a rupture in the text of the film. In *Yo soy la felicidad de este mundo* I originally wanted to show an ejaculation in the central scene of masturbation, but unfortunately the actor proved unprepared for that eventuality.

Q: There's a camera movement that is very typical of your films: the 360-degree pan or dolly. Often you begin with one boy in bed and when the camera returns to its original position his place has been taken by another. What is the special attraction that this camera movement has for you? How do you use it in your films?

A: The camera is just another character in my cinema and I think it adopts an amorous relationship with the other characters. Another way of putting it would be to say that the director caresses the bodies of the actors with the camera. Along the same lines, I would also suggest that the camera's relationship with the actors is erotic, a kind of dance of seduction that envelops them in its movement.

Q: What is the nature of your work with the actors? Do you rehearse a great deal? Is there space for improvisation on the set or do you follow the script as written?

A: Even though they have very little dialogue, my scripts are very specific. I like them to be literary and to be a pleasure to read. This literary quality of the script thus becomes part of my work with the actors, as it gives them guidance when they come to read it. On the other hand, I wait until the shoot for rehearsal. And as I said earlier, the role of location is vital in producing the performance. As there is no rehearsal process before filming, there is indeed a possibility of improvisation, or of artistic construction between director and actor on set. This can lead to surprises, especially as my actors are often dancers, who are used to expressing themselves through their body. This kind of actorly movement also contributes to the construction of character on set.

Q: Your plots generally exhibit a certain tendency toward fragmentation. For example, *Rabioso sol, rabioso cielo* is made up of two separate halves: a modern urban section and a rural mythical section (included only in the DVD version of the film and not seen in the theatrical release). Likewise, the two melancholy love stories of *Yo soy la felicidad de este mundo* are separated by the long dance cum sex sequence that seems to have come out of another film. Are you conscious of this will to fragmentation in your practice as a screenwriter or is it determined by other factors such as your vision of the film as a whole?

A: My narratives are intended to be disruptive, fragmented. I conceived *Yo soy la felicidad de este mundo* as a triptych in three scenes where the same elements are reflected and repeated in each part. The sections are thus like [the Surrealist technique of] communicating vessels. I try to free myself from the clarity of plot and invent a different structure in which I abandon the established grammar of film for a model of cinema as a state of mind. But I'm still very much aware of the need to please an audience. I like to cite Pasolini when he said that the only thing he asks of the audience is a level of commitment equal to that of the director. However, the aim of my films is nonetheless to produce pleasure and enjoyment in the spectator.

Q: You've overcome a great challenge and had great success in doing something quite unheard of in Mexico: making a whole corpus of experimental cinema with a queer theme. Can you talk about the production process that has permitted this and your relationship with your partner and producer Roberto Fiesco?

A: For ten years, from my first feature to my third, Roberto and I were able to produce continuously without any interruption. However, after that time, and in spite of our track record, we were unable to go ahead with our next project, which ironically enough was a genre movie, a film noir. So we decided instead to make *Yo soy la felicidad de este mundo* on a very low budget in just fourteen days of shooting. It seems more difficult now to make our kind of films, although the noir will start production soon with the working title *Rencor tatuado* ("Tattooed Rancor"). As for queer themes, the attitude to sexual diversity in Mexico has changed radically since the time when we started working, and there is now no rejection of gay movies or homosexuality per se. Open homophobia is no longer acceptable in Mexico, although of course it still exists in a more veiled fashion. Stigmatization and labeling are still around, however, and there is no equality.

Q: You're very active on Instagram (as instagram.rabioso) and Twitter (as @julianherper). Are social media important to you? Do you see them as a way to keep in contact with your audience?

A: I see social media as a space where I can express myself and enjoy the freedom to say what I want. I particularly like it because I am actually a fun person and, as my movies are serious, nobody believes that's the way I am. I think that writing a tweet can be like making a short film. And it is indeed a good way of keeping in contact with the audience. I don't react badly when people say negative things about my films on Twitter. Rather, I treat social media as a kind of exquisite vulgarity. Perhaps a goal would be for me to make films in the same way that I tweet.

Q: To conclude, *Yo soy la felicidad de este mundo* has for the first time in your cinema a protagonist who is a filmmaker and, indeed, a gay auteur in the process of making a new feature. The film thus offers an ironic and self-reflexive commentary on your own professional situation. Do you have any final comments on your practice in this meta perspective?

A: In my earlier films the main characters were seeking a sense of completeness in the other person whom they desired. In the new film, *Yo soy la felicidad de este mundo*, this theme is transformed and the love affairs of the protagonist, a filmmaker, mask a profound inability to relate to others. Although the film is certainly not autobiographical, it does express an ironic approach to my own position as a director. One important thing is that it is the male prostitute, the director's second lover, who is the most generous and good-hearted character in the film. He's also a kind of innocent, in that he is amazed to meet a filmmaker whose work he is familiar with. With this character, I go back once more to classic Mexican cinema of the Golden Age, where the [female] prostitute is seen as the innocent victim of violence. But the difference here is that the sex worker does not remain a victim and indeed walks out on his abusive lover. Some people have told me that they recognized the actor from escort websites in Mexico, but that's not true; of course, in real life he's an actor not a prostitute! In fact, rather than identifying myself with the director in the film, I identify more with the sex worker, as you could say that a film director is a bit like a rent boy.

This interview has been translated and condensed from the original Spanish. My thanks to Daniela Michel, founding director of the Morelia International Film Festival, for her kind invitation.

WORKS CITED

Aguilar, Arturo. 2011. "Morir de pie." *Letras Libres*, April 25.

———. 2015. "La vuelta al cine negro." *Gatopardo*, October. http://gatopardo.com/EstilosHomeGP.php?Id=1364.

Aguilar Ascencio, Patricia. 2001. "La construcción imaginaria del estereotipo de la homosexualidad: Discurso cinematográfico y subjetividad." PhD diss., UAM Xochimilco.

Aguirre, Mariana. 2012. "Interview with Critic, Curator, and Art Historian Cuauhtémoc Medina." *Art 21 Magazine*, August 3. http://blog.art21.org/2012/08/03/interview-with-critic-curator-and-art-historian-cuauhtemoc-medina-part-1/#.VvauHTbpfa5.

"Al final del arcoiris: Temporada 2; Verde." 2009. *YouTube*, December 3. www.youtube.com/watch?v=yHvHwLLE6Jw.

Andreu, Isabel. 2014. "El Viaje de Morgana." Interview with Flavio Florencio and Morganna. *Noir Magazine*, July 20.

Ayala Blanco, Jorge. 2004a. "Festival Mix y los demás: Pálpitos." *El Financiero*, May 24, section Cultural.

———. 2004b. "Festival Mix: Zeugmas." *El Financiero*, May 31, section Cultural.

———. 2006. "X Festival Mix: Resguardando." *El Financiero*, May 29, section Cultural.

———. 2008. "Festival Mix Platino: Orientando." *El Financiero*, June 2, section Cultural.

Beasley-Murray, Jon. 2011. *Posthegemony*. Minneapolis: University of Minnesota Press.

Bloch, Catherine. 2013. "Pluralidad de talento y entrega total." *Cine Toma* 29 (5): 66–69.

Bonfil, Carlos. 2004. "Festival Mix de diversidad sexual." *La Jornada*, May 30, section Espectáculos.

———. 2006. "Feliz Mix 2006." *La Jornada*, May 29, section Espectáculos.

———. 2007. "Mix Persona." *La Jornada*, May 27, section Espectáculos.

———. 2011. "Una familia muy normal." *La Jornada*, May 15, section Espectáculos.

———. 2013. "Quebranto." *La Jornada*, June 2. www.jornada.unam.mx/2013/06/02/opinion/a10a1esp.

———. 2014. "Mix México 2014." *La Jornada*, June 8, section Espectáculos.

———. 2015a. "Foro de La Cineteca: Made in Bangkok." *La Jornada*, July 8, section Espectáculos.

———. 2015b. "Morelia 2015: Los encuentros venturosos." *La Jornada*, November 1. www.jornada.unam.mx/2015/11/01/opinion/a09a1esp.

Bourdieu, Pierre. 1994. *The Field of Cultural Production*. New York: Columbia University Press.

———. 1996. *The Rules of Art*. Cambridge: Polity.

Bretón, Alexandra. 2004. "Se cubre Cineteca con diversidad." *El Universal*, May 21.

Brooks, Chad. 2015. "What Is Entrepreneurship." *Business News Daily*, January 5. www.businessnewsdaily.com/2642-entrepreneurship.html.

Bruciaga, Wenceslao. 2013. "Luces, cámara y ¡acción gay!" *Time Out México*, May.

Bustamente, Maza. 2013. "El Mix dominante." *Milenio*, May 25, section El Sexódromo.

Cadena Tres Espectáculos. 2015. "Juan Manuel Bernal reacciona ante rumores de sus preferencias sexuales." *YouTube*, February 4. www.youtube.com/watch?v=K30sI_Zrdls&list=PLf2gHQuMkWt_5LcVKId19hv4jWTX2RFHP&index=1.

"*Capadocia*, historia de una cárcel de mujeres, se estrena por HBO." 2008. *La Jornada*, February 22.

Cárdenas Ochoa, Alejandro. 2005. "Le apuestan a la diversidad sexual." *El Universal*, May 15, section Espectáculos.

———. 2006. "Exploran la sexualidad aplauden la diversidad." *El Universal*, May 19, section Por Fin.

———. 2008. "Enciende Bramadero los ánimos." *Bramadero Blogspot*, January 28. http://bramadero.blogspot.com/2008_01_01_archive.html.

———. 2013. "Quebranto." *La Jornada*, June 2, section Espectáculos.

Carrillo, Héctor. 2002. *The Night Is Young: Sexuality in Mexico in the Time of AIDS*. Chicago: University of Chicago Press.

Casarin, Susana. 2012. *Realidades y deseos*. Mexico City: Artes de México.

Castañeda, Iván. 2016. "Las Aparicio admiran a Kate del Castillo." *Milenio*, February 23. www.milenio.com/hey/cine/Las_Aparicio_admiran_a_Kate-historias_de_mujeres_chingonas-Erendira_Ibarra_0_688731155.html.

Castro Ricalde, María de la Cruz. 2015. "Lesbians Made in Mexico: Sexual Diversity and Transnational Fluxes." In *Despite All Adversities: Spanish-American Queer Cinema*, edited by Andrés Lema-Hincapié and Debra A. Castillo, 203–18. Albany: State University of New York Press.

Charlois Allende, Adrien, ed. 2011. *Telenovelas en México: Nuestras íntimas extrañas*. Mexico City: Grupo Delphi.

"Cine Teresa, deja el porno por el arte." 2013. *24 Horas*, March 20.

Cinépolis. 2014. Synopsis, *Cuatro lunas*. http://www.cinepolis.com/pelicula/cuatro-lunes.

Cobian, Jaime. 2013. *Los jotos: Cronología y diccionario*. Guadalajara: Promoteo.

Conarte. 2014. Synopsis, *Carmín tropical*. www.conarte.org.mx/evento/carmin-tropical.

Cordero Rentería, Rosa Angélica. 2005. "Jaime Humberto Hermosillo: Algunas transgresiones en la sexualidad femenina." PhD diss., UNAM.

Cruz Bárcenas, Arturo. 2011. "La mediocridad de la industria, el verdadero reto para un actor." *La Jornada*, April 26.

———. 2015. "El corto es territorio libre; es el hermano bastardo del cine." *La Jornada*, September 12.

Cueva, Álvaro. 2010. "Tele marica." In *México se escribe con J*, edited by Michael Schuessler and Miguel Capistrán, 177–83. Mexico City: Editorial Planeta Mexicana.

———. 2011. "Confesiones de un crítico de telenovelas." In *Telenovelas en México: Nuestras íntimas extrañas*, edited by Adrien Charlois Allende, 9–44. Mexico City: Grupo Delphi.

De Baecque, Antoine. 2003. *La Cinéphilie*. Paris: Fayard.

De la Fuente, Anna Marie. 2006a. "'Brokeback' Faces Latin Culture Test." *Variety*, March 12. http://variety.com/2006/film/markets-festivals/brokeback-faces-latin-culture-test-1117939565/.

———. 2006b. "Guadalajara Changes Guard." *Variety*, March 20, A1.

De la Mora, Sergio. 2006. *Cinemachismo: Masculinities and Sexuality in Mexican Film*. Austin: University of Texas Press.

Díaz Mendiburo, Aarón. 2004. *Los hijos homoeróticos de Jaime Humberto Hermosillo*. Mexico City: Plaza y Valdés.

Divers Magazine. 2015. December.

"*El sexo debil*—A Daring, Revealing Mexican Series Starring Mauricio Ochmann, Itati Cantoral, and Khotan Fernández—Debuts July 6 on NBC Universo." 2015. *NBC Universal*, June 25.

Esparza, Enrique. 2011. "Documentalistas presentan sus propuestas." *Informador.com.mx*, April 29.

Farber, Stephen. 2014. "'I Am Happiness on Earth': Film Review." *Hollywood Reporter*, August 14.

"Festival Mix 2006 presentará un filme dedicado a la vida gay." 2006. *La Crónica*, May 17, section Espectáculos.

FICG. 2011. Synopsis of *Morir de pie*.

———. 2013. Synopsis of *Quebranto*.

Fiesco, Roberto. 2009. "Sexxxcuestro: Una experiencia pornográfica; Entrevista a Laars Robledo y Summer Gandolf." *Cine Toma* 5 (1): 34–36.

———. 2014. "Buscan evitar condena social a transgéneros con documental." *Milenio*, January 12. www.milenio.com/cultura/Buscan-evitar-condena-transgeneros-documental_0_225577481.html.

Foster, David William. 2010. "Documenting Queer, Queer Documentary." *Revista Canadiense de Estudios Hispánicos* 35 (1): 105–19.

García Hernández, Arturo. 1999. "*La vida en el espejo* será ahora la nueva *Mirada del hombre.*" *La Jornada*, June 27.

García Tsao, Leonardo. 2013. "Dos documentales, dos." *La Jornada*, March 16, section Espectáculos.

González, Jesús Isaac. 2016. "Reinventar a Azteca." Interview with Benjamín Salinas. *Líderes*, 44–50. March.

Gracida, Ysabel. 2005. "Festival Mix." *El Universal*, May 20, section Cultura.

Guadarrama Rico, Luis A. 2014. "Imperativos machistas." *Milenio*, December 8.

Gutiérrez García, Gabriel. 2005. "9 años de festín visual sexual." *La Guía de México: Tiempo Libre*, May 19.

Hecht, John. 2006. "Buyers See New Relevance for Guadalajara Film Mart." *Hollywood Reporter*, 20. March 31.

Hernández, Julián. 2016. "El cielo dividido (2006) secuencias borradas uno." Vimeo. https://vimeo.com/151670911.

"Historia de amor cierra Festival Mix." 2005. *El Universal*, May 27, section Cultura.

Holmlund, Chris, and Cynthia Fuchs. 1997. *Between the Sheets, in the Streets: Queer, Lesbian, Gay Documentary*. Minneapolis: University of Minnesota Press.

Huerta, César. 2007. "Muestran cine gay y sobre violencia." *El Universal*, May 18, section Espectáculos.

———. 2015a. "Made in Bangkok ante el tribunal del DF." *El Universal*, October 30, section Espectáculos.

———. 2015b. "Vida de soprano transgénero triunfa en el GIFF." *El Universal*, July 27, section Espectáculos.

"I Am Happiness on Earth." 2014. *Naked Guys in Movies*. www.nakedguysinmovies.com/content/alan-ram-rez-gabino-rodr-guez-iv-n-lvarez-emilio-von-sternerfels-hugo-catal-n-and-gerardo.

IMDb. 2012. Synopsis, *Everybody's Got Somebody . . . Not Me.*

———. 2014. Synopsis, *Four Moons*.

"Jacaranda Correa quiere *Morir de pie* en homocinefilus." 2011. *Homecinefilus*, March 30.

Jara, Rubén. 2011. "Telenovela y rating." In *Telenovelas en México: Nuestras íntimas extrañas*, edited by Adrien Charlois Allende, 121–60. Mexico City: Grupo Delphi.

Jenkins, Henry. 2006. *Convergence Culture*. New York: New York University Press.

———. 2013. *Spreadable Media*. New York: New York University Press.

Juett, JoAnne C., and David M. Jones, eds. 2010. *Coming Out to the Mainstream: New Queer Cinema in the 21st Century*. Newcastle: Cambridge Scholars Press.

"Julián Hernández: Homoerotismo en la gran pantalla." 2012. *Zonadiversa*. www.zonadiversa.com/entrevistas/23-julian-hernandez-homoerotismo-en-la-gran-pantalla.

Koehler, Robert. 2009. "Homegrown Fiesta: Bumper Crop of Mexican Features." *Variety*, A5. March 16.

"La vida en el espejo." 2016. Wikipedia. https://es.wikipedia.org/wiki/La_vida_en_el_espejo.

Legnar, Harry. 2014. "Armando a Silva Baena: ¡6 años a color! [interview]." *Homos*, no. 18 (November): 17–38.

Lema-Hincapié, Andrés, and Debra A. Castillo, eds. 2015. *Despite All Adversities: Spanish-American Queer Cinema*. Albany: State University of New York Press.

"Llega a las salas cinematográficas El 9° Festival de diversidad sexual en cine y video." 2005. *Dónde*, May, section Cine.

López, Sergio Raúl. 2009. "'Uno es platillo y el otro la receta': Entrevista al Diablo, director de la productora Mecos Films." *Cine Toma* 5 (1): 32–34.

López Pumarejo, Tomás. 2013. "Las televisiones de EEUU: Cómo se adaptan al cambio." In *Convergencia y transmedialidad*, edited by Lorenzo Vilches, 307–17. Barcelona: Gedisa.

Maclaird, Misha. 2013. *Aesthetics and Politics in the Mexican Film Industry*. New York: Palgrave Macmillan.

Maitra, Ani. 2015. "Touching Hearts: The Uncertain but Strategic Politics of Kashish 2015." *Film Quarterly* 69 (2): 60–65.

Marquet, Antonio. 2006. *El crepúsculo de heterolandia: Mester de jotería*. Mexico City: UAM Azcapotzalco.

Martínez, Pablo. 2014. "Descubrimos que Juan Manuel Bernal tiene desde hace 3 años un amor secreto." *TVNotas*, May 13. http://archive.is/1XF9Y.

"Mexico Fetes Cuarón's Oscars, but Filmmakers Keep Feet on Ground." 2014. *States News Service*, March 3.

Mix. 1998. "Segundo festival de diversidad sexual en cine y video." Film festival, March 6–27.

———. 1999. "Tercer festival de diversidad sexual en cine y video." Film festival, April 15–30.

———. 2000. "Cuarto festival de diversidad sexual en cine y video." Film festival, April 5–20.

———. 2001. "Quinto festival de diversidad sexual en cine y video." Film festival, May 2–20.

———. 2002. "Sexto festival de diversidad sexual en cine y video." Film festival, May 2–9.

———. 2003. "Séptimo festival de diversidad sexual en cine y video." Film festival, May 10–18.

———. 2004. "Octavo festival de diversidad sexual en cine y video." Film festival, May.

———. 2005. "Noveno festival de diversidad sexual en cine y video." Film festival, May.

———. 2006. "Décimo festival de diversidad sexual en cine y video." Film festival, May.

———. 2007. "Undécimo festival de diversidad sexual en cine y video." Film festival, May 22–June 6.

———. 2008. "Doceavo festival de diversidad sexual en cine y video." Film festival, May.

———. 2009. "Décimo tercero festival de diversidad sexual en cine y video." Film festival, May.

———. 2010. "Decimocuarto festival de diversidad sexual en cine y video." Film festival, May 25–June 13.

———. 2011. "Decimoquinto festival de diversidad sexual en cine y video." Film festival, May.

———. 2013. "Decimoséptimo festival de diversidad sexual en cine y video." Film festival, May–June.

———. 2015. "Decimonoveno festival de diversidad sexual en cine y video." Film festival.

Monsiváis, Carlos. 2010. *Qué se abra esa puerta: Crónicas y ensayos sobre la diversidad social*. Mexico City: Paidós.

Montaño Garfias, Ericka. 2009. "Susana Casarín [*sic*] explora fragmentos de la vida de cuatro mujeres en el cuerpo de un hombre." *La Jornada*, March 18. www.jornada.unam.mx/2009/03/18/cultura/a05n2cul.

Moscatel, Susana. 2016. "Una buena para Tv Azteca." *Milenio*, 48. April 5.

Muñoz Tinoco, Óscar. 2013. "'Quebranto': La historia de un hombre que quiso ser mujer." *Filmeweb*. http://filmeweb.com.mx/blog/tag/roberto-fiesco/.

Olivares, Juan José. 2004. "El Festival Mix, historias profundas e intensas, más allá de la banalidad." *La Jornada*, May 19, section Espectáculos.

Orozco, Guillermo. 2011. "Entre espectáculo, mercado y política: La telenovela mexicana en más de cinco décadas." In *Telenovelas en México: Nuestras íntimas extrañas*, edited by Adrien Charlois Allende, 181–218. Mexico City: Grupo Delphi.

O'Toole, Gavin. 2010. *The Reinvention of Mexico: National Ideology in a Neoliberal Era*. Liverpool: Liverpool University Press.

Pérez, Javier. 2005. "Sexualidad múltiple." *El Universal*, May 20, section Por Fin.

Pérez Mancilla, Ulises. 2007a. "El cielo dividido." El cielo dividido. http://elcielodividido.blogspot.com/.

———. 2007b. "Harto rabioso, parte 2." *Rabioso sol, rabioso cielo*, July 31.

Pérez-Sánchez, Gemma. 2010. "Transnational Conversations in Migration, Queer, and Transgender Studies: Multimedia Storyspaces." *Revista Canadiense de Estudios Hispánicos* 35 (1): 163–84.

Piñón, Juan. 2011. "The Unexplored Challenges of Television Distribution: The Case of Azteca America." *Journal of Television and New Media* 12 (1): 66–99.

———. 2014. "Webnovelas: Branding Interactivity in Hispanic TV." *Popular Communication* 12 (3): 123–38.

Podalsky, Laura. 2011. "Landscapes of Subjectivity in Contemporary Mexican Cinema." *New Cinemas: Journal of Contemporary Film* 9 (2 and 3): 161–82.

Preciado, Beatriz. 2002. *Manifiesto contrasexual*. Barcelona: Anagrama.

"*Quebranto*, la nostalgia del pasado frente a un duro presente." 2013. *IMCINE*, February 22.

Quintero Murguía, Rodrigo. 2015. *La ruta del arcoíris en la sultana: Historia del movimiento LGBT en Nuevo León*. Monterrey: Fondo.

Rich, B. Ruby. 2006. "The New Homosexual Film Festivals." *GLQ: A Journal of Lesbian and Gay Studies* 12 (4): 620–25.

———. 2013. *New Queer Cinema: The Director's Cut*. Durham, NC: Duke University Press.

Rocha, Alberto. 2007. "Cambio de look." *Dónde*, May, section Cine.

Rodríguez, Irayda. 2015. "José María Yazpik sin rodeos." *Capital55*, April, 49–53.

Ruiz, Stefan. 2012. *Factory of Dreams*. New York: Aperture.

"Sabrina le hace el feo a su hija lesbiana." 2015. *TVyNovelas*, April 6.

Salgado, Ivett. 2013. "Plasma Roberto Fiesco una historia diferente." *Milenio*, December 28.

———. 2016. "Abordan la diversidad del amor en serie de Juanga." *Milenio*, February 25. www.milenio.com/hey/television/Abordan-diversidad-amor-serie-Juanga_0_689931011.html.

Sánchez, Julieta, and Javier Pérez. 2011. "Todos los personajes buscan una familia." *Cine Toma* 14 (3): 68–69.

Sánchez Prado, Ignacio M. 2014. *Screening Neoliberalism: Transforming Mexican Cinema, 1988–2012*. Nashville: Vanderbilt University Press.

Schuessler, Michael, and Miguel Capistrán, eds. 2010. *México se escribe con J: Una historia de la cultura gay*. Mexico City: Temas de hoy.

Schulz-Cruz, Bernard. 2008. *Imágenes gay en el cine mexicano: Tres décadas de joterío, 1970–1999*. Mexico City: Fontamara.

"¡Se acabaron las telenovelas de las 4 de la tarde en Televisa!" 2016. *TVNotas*, March 14.

Smith, Paul Julian. 2014. *Mexican Screen Fiction: Between Cinema and Television*. Cambridge: Polity.

———. 2015. "Screenings: The Caress of the Camera in the Cinema of Julián Hernández [interview]." *Film Quarterly* 3 (68): 85–92.

Solís, Juan. 2002. "El cine, otra pasión de Luis Zapata." *El Universal*, May 15, section F.

———. 2003. "El Mix es un festival provocativo: Castelán." *El Universal*, May 12, section Cultura.

———. 2004. "La mirada alterna que cuestiona e incomoda." *El Universal*, May 20, section Cultural.

Straayer, Chris. 1997. "Transgender Mirrors: Queering Sexual Difference." In *Between the Sheets, in the Streets: Queer/Lesbian/Gay Documentary*, edited by Chris Holmlund and Cynthia Fuchs, 207–33. Minneapolis: University of Minnesota Press.

"'Sube de tono' El Festival de Morelia estrenan [*sic*] 'Bramadero.'" 2008. "BRAMADERO." *Bramadero Blog*, January 14. http://bramadero.blogspot.com/2008/01/sube-de-tono-el-festival-de-moreliasube.html.

Subero, Gustavo. 2009. "Gay Mexican Pornography at the Intersection of Ethnic and National Identity in Jorge Diestra's La Putiza." *Sexuality and Culture* 3 (14): 217–33.

———. 2012. "Gay Male Pornography as Latin American His/Toriography." In *LGBT Transnational Identity within Media: Post-Colonial, Post-Queer*, edited by Christopher Pullen, 213–30. Basingstoke: Palgrave Macmillan.

———. 2013. *Queer Masculinities in Latin American Cinema: Male Bodies and Narrative Representations*. London: I. B. Tauris.

"Synopsis of *El cielo dividido*." 2006. *El séptimo arte*. www.elseptimoarte.net/peliculas/el-cielo-dividido-808.html.

"The Two Mexicos." 2015. *Economist*, September 19–25.

Torrano, María de Lourdes. 2015. "Un festival que no descrimina." *Más por Más DF*, February 6.

Tovar, Luis. 2006. "La moneda y sus caras." *La Jornada*, May 28, section Jornada semanal.

———. 2013. "Sin la frente marchita." *La Jornada*, December 15. www.jornada.unam.mx/2013/12/15/sem-tovar.html.

Tovar Velarde, Sergio. 2015. *Cuatro Lunas* (novelization of feature film). Mexico City: Planeta.

"Transita *Morir de* pie de las revoluciones sociales a las íntimas." 2011. *IMCINE*, March 29.

"Un festival que no descrimina." 2015. *Más por Más DF*, February 6, section First.

Van Hoeij, Boyd. 2014a. "'Carmin Tropical': Morelia Review." *Hollywood Reporter*, October 31. www.hollywoodreporter.com/review/carmin-tropical-morelia-review-745146.

———. 2014b. "'Four Moons' ('Cuatro Lunas'): Morelia Review." *Hollywood Reporter*, November 3. www.hollywoodreporter.com/review/four-moons-cuatrolunas-morelia-745579.

Vargas, Susana, ed. 2015. *Mujercitos*. Mexico City: RM.

Williams, Linda. 1989. *Hard Core: Power, Pleasure, and the Frenzy of the Visible*. Berkeley: University of California Press.

"Yo soy la felicidad de este mundo." 2014. FilmAffinity. www.filmaffinity.com/es/film433779.html.

Young, James. 2014. "Mexican Stronghold for Talent: Iñárritu's 'Birdman' Leads Strong Selection of Homegrown Fare." *Variety*, October 14.

Záizar Arellano, Christyan. 2005. "Muestra en cine y video sobre la diversidad sexual." *Excelsior*, May 21, section Espectáculos.

Zamora, Fernando. 2015. "¡Knock Out! [review of *Carmín Tropical*]." *Milenio: Laberinto*, October 17.

INDEX

Page references with an *f* are figures.

activism, 1, 3, 88; Mix Festival, 11 (*See also* Mix Festival)
activos, 37, 43
actors, working with, 147
adult films, 26. *See also* pornography
AEBN (Adult Entertainment Broadcast Network), 30
Alarma!, 64
Alerta!, 65
Al final del arcoíris (At the End of the Rainbow), 19–24, 44
Alliance for Cultural Rights of Sexual Diversity and Non-Discrimination, 15
Almodóvar, Pedro, 48
Altman, Dennis, 12
ambulantes (street vendors), 76
Anderson, Harriet, 87
Las Aparicio (Cadena 3), 110, 138, 139
Aquí no hay quien viva (No One Can Live Here), 107
Argos (telenovelas), 107–135, 138; *Capadocia,* 128–134; format of telenovelas, 110; Life in the Mirror *(La vida en el espejo),* 108–110, 113*f,* 114–122; The Weaker Sex *(El sexo débil),* 120*f,* 121–127; A Woman's Look *(Mirada de mujer),* 110, 111
Arroyo, Fernando, 45, 46
art, rules of (Bourdieu), 58
art cinema, 1, 31, 86, 104. *See also* films
Artes de México magazine, 65
A Summer Dress *(Une Robe d'été),* 9
A Thousand Clouds of Peace Encircle the Sky, Love, Your Being Love Will Never End *(Mil nubes de paz cercan el cielo, amor, jamás acabarás de ser amor),* 90, 143, 144, 145
Atmósfera, 55
At the End of the Rainbow *(Al final del arcoíris),* 19–24, 30, 44
audiences for telenovelas, 110
audiovisuals, economics of queer, 8

A Woman's Look *(Mirada de mujer),* 110, 111
Ayala Blanco, Jorge, 16, 17
Azteca, 2, 123, 131

ballet, 51
Bausch, Pina, 48
Bazin, André, 86
Beasley-Murray, Jon, 88, 94
The Beating *(La putiza),* 27
Bell, Monna, 55
Bernal, Juan Manuel, 100, 130, 133
Between the Sheets, in the Streets: Queer, Lesbian, Gay Documentary (Holmlund/Fuchs), 62
Birdman, 36
bisexuales, 38
bisexuality, 50
Bonelli, Coral, 71–77, 80, 82, 84
Bonfil, Carlos, 5, 15, 16, 73, 79, 137
Bourdieu, Pierre, 58
box office, 35
Bracho, Julian, 145
Bramadero, 28, 52–54, 56, 146
Brokeback Mountain, 35
Broken Sky *(El cielo dividido),* 16, 33–36, 40*f,* 42, 143; need for love in, 43; Podalsky, Laura, 45; synopsis, 40, 41; in the university environment, 46
Bruciaga, Wenceslao, 17, 18

Cadena 3, 110, 121
El callejón de los milagros (Midaq Alley), 97
cameras, 147
Cantú, Alejandro, 48
Capadocia, 100, 128*f,* 129–134, 138; synopsis, 128
Capital55 magazine, 111, 112
Cárdenas, Nancy, 4
Carmín tropical (Tropical Lipstick), 86, 101*f,* 101–104, 106; synopsis, 101
Carrillo, Héctor, 36–39, 43, 45, 59
Casarin, Susana, 65, 66, 69, 72, 82

CasAzul, 134
Castelán, Arturo, 11, 13–15, 24, 34, 71
Catalán, Hugo, 47, 58
The Celluloid Closet, 14
characters, queer, 3
El cielo dividido (Broken Sky), 16, 33–36, 40*f*, 42, 143; need for love in, 43; Podalsky, Laura, 45; synopsis, 40, 41; in the university environment, 46
Cinemachismo: Masculinities and Sexuality in Mexican Film (de la Mora), 38
cinéma vérité, homosexualization of, 23
cinephilia, 86, 87, 90, 93, 99. *See also* Everyone Has Someone But Me *(Todo el mundo tiene a alguien menos yo)*
Cinépolis Diana, 15, 17
Coming Out to the Mainstream (Juett/Jones), 2
Correa, Jacaranda, 67, 68, 70, 71. *See also* To Die Standing Up *(Morir de pie)*
Corrupción mexicana, 27, 28, 29, 30
counterhegemony, 88
Crawford, Joan, 102
critics, telenovelas, 109
Cuarón, Alfonso, 3, 4, 36, 115
Cuatro lunas (Four Moons), 86, 94*f*, 96–100, 104–106, 111; synopsis, 95
Cuba, 2
Cuban Revolution, 69
CUEC (University Center of Cinema Studies), 26, 144
Cueva, Álvaro, 108, 109, 111
cult of Our Lady of Guadalupe, 39
cultures, Mexico, 37

Daeva, Naian, 90
Damiens, Antoine, 12
dance, 51
David, 49
de Baecque, Antoine, 86, 87
De la Fuente, Anna Marie, 34
de la Mora, Sergio, 38, 39, 49
de la Reguera, Ana, 131, 138
del Castillo, Kate, 23
del Río, Dolores, 64
del Toro, Guillermo, 36
demographics for telenovelas, 110
Derbez, Eugenio, 107
DFÑOS (MexicoCitizens), 23
Díaz Mendiburo, Aarón, 5
Disrupted (Quebranto), 70*f*, 73–77, 79, 80, 83

distribution, 8, 58
Divers magazine, 4
documentaries, 1, 14; *The Celluloid Closet*, 14; Fiesco, Roberto, 145; *Quebranto*, 15; techniques in fiction, 101, 102 (*See also* Tropical Lipstick *[Carmín tropical]*); The Transformation of Film into Music *(La transformación del cine en música)*, 145; transgender, 61–84 (*See also* transgender documentaries); Young Man at the Bar *(Muchacho en la barra)*, 90
"Documenting Queer, Queer Documentary" (Foster), 64

Easter eggs, 49
economics of queer audiovisuals, 8
El Economista, 16
Egelhaaf, Gustavo, 99
Eimbcke, Fernando, 44
entrepreneurship, 7, 8
The Erect Cock *(La Verga Parada)*, 27
eroticism, 138
Escalante, Amat, 5, 105
Everyone Has Someone But Me *(Todo el mundo tiene a alguien menos yo)*, 85, 86, 89*f*, 91–94, 104, 106; synopsis, 89
explicit sex in films, 50

Facebook, 31
Factory of Dreams, 139
Faites comme chez vous (Make Yourself at Home), 107
Falikov, Tamara, 12
feature films. *See* films
Félix, María, 102
femenina, 38
Fernández, Emilio "El Indio," 144
Fernández, Joaquín, 75
Festival of Sexual Diversity in Film and Video. *See* Mix Festival (2014)
festivals, 1, 3, 139; advances for gays, 16; audiences, 18; films, 12; Freiburg, Germany, 13; global locations of Mix, 14; Guadalajara International Film Festival, 34, 67, 71, 78; Guanajuato Film Festival, 78; history of, 17; homophobia, 15, 16; India (Kashish), 12, 13; Los Angeles, California, 13; Mexican Human Rights Film Festival, 78; Mix Festivals, 7, 8, 9, 10*f*, 11, 13, 30, 34, 87; Morelia International Film Festival, 2, 35,

72, 78, 104, 149; The New Homosexual Film Festivals, 34; panels, 12; Society of Cinema and Media Studies, 12; Zanzibar International Film Festival, 78
fictions: documentaries and, 63; mix of reality and, 73
Fiesco, Roberto, 2, 3, 15, 16, 26, 47, 59, 143; *David*, 49; Disrupted *(Quebranto)*, 70*f*, 72–77; documentaries, 145; *La Jornada*, 52; as producer, 33
Filmeweb.net, 72
Film Quarterly, 12
films, 1; adult, 26 (*See also* pornography); The Beating *(La putiza)*, 27; *Birdman*, 36; box office, 35; *Bramadero*, 146; *Brokeback Mountain*, 35; Broken Sky *(El cielo dividido)*, 16, 33–36, 40*f*, 42–46, 143; *Capadocia*, 138; *Corrupción mexicana*, 27–30; The Dark Springs *(Las oscuras primaveras)*, 111; documentaries, 14; explicit sex in, 50; *Factory of Dreams*, 139; festivals, 3, 12 (*See also* festivals); Four Moons *(Cuatro lunas)*, 111; French New Wave, 91, 93; I Am Happiness on Earth *(Yo soy la felicidad de este mundo)*, 33, 34, 40, 46*f*, 51, 90, 143–149; I Promise You Anarchy *(Te prometo anarquía)*, 137–140; *La Jetée*, 9; Ladies Dress Designer *(Modisto de señoras)*, 140; lesbians, 31; mainstream movies, 85–106 (*See also* mainstream movies); Mecos, 41 (*See* Mecos); *Mexican Men*, 140; Midaq Alley *(El callejón de los milagros)*, 97; Northless *(Norteado)*, 104; Raging Sun, Raging Sky *(Rabioso sol, rabioso cielo)*, 14, 33, 35, 52, 72, 144, 145, 147; *Rear Window*, 87, 90; Sex, Shame, and Tears *(Sexo, amor y lágrimas)*, 114; Talk to Her *(Hable con ella)*, 48; A Thousand Clouds of Peace Encircle the Sky, Love, Your Being Love Will Never End *(Mil nubes de paz cercan el cielo, amor, jamás acabarás de ser amor)*, 90, 143, 144, 145; transgender documentaries (*See* transgender documentaries); *Veronika Voss*, 48, 144; *Vestido*, 9, 18; XXX Kidnap *(SeXXXcuestro)*, 26, 28, 29, 30
Florencio, Flavio, 77*f*, 79, 81, 82
fluidity of sexuality, 13
Fondo Nacional para la Cultura y las Artes (National Fund for Culture and the Arts), 13
Fons, Jorge, 71, 72, 74, 97
FOPROCINE, 78

Fortissimo Films, 35
Foster, David William, 64
Four Moons *(Cuatro lunas)*, 86, 94*f*, 95–100, 104–106, 111; synopsis, 95
free aggregation sites, 8
Freiburg, Germany, 13
French New Wave films, 91, 93
Fuchs, Cynthia, 62, 63, 70
Fuentes, Raúl, 85, 86, 89*f*, 91–94, 104, 106

Gabriel, Juan, 140
Gaitán, Paulina, 9, 15
Garcés, Mauricio, 140
García Bernal, Gael, 39
García Tsao, Leonardo, 73
Gavaldón, Roberto, 145
gay: Bonelli, Coral, 82 (*See also* Bonelli, Coral); definition of, 37, 38; dramas, 7; identities, 62; subcultures, 42
gender as performance, 62. *See also* transgender documentaries
Golden Age cinema, 43, 45, 53, 59, 61, 102, 145, 149
González Iñárritu, Alejandro, 36
González Norvind, Naian. *See* Daeva, Naian
Guadalajara International Film Festival, 34, 67, 71, 78
Guanajuato Film Festival, 78
Guau, 61
Guevara, Che, 2, 68

Hable con ella (Talk to Her), 48
Haynes, Todd, 15
HBO Latin America, 128–133
Hecht, John, 35
hegemony, 88
Hermosillo, Jaime Humberto, 5, 33
Hernández, Julián, 1, 2, 3, 8, 11, 14, 31, 71, 72, 83, 85, 111, 137, 138, 140; actors, working with, 147; art cinema, 33–59; *Atmósfera*, 55; *Bramadero*, 28, 52–56, 146; Broken Sky *(El cielo dividido)*, 16, 33–36, 40*f*, 42–46, 143; cameras, 147; *David*, 49; Disrupted *(Quebranto)*, 71; distribution, 58; Easter eggs, 49; Four Moons *(Cuatro lunas)*, 111; I Am Happiness on Earth *(Yo soy la felicidad de este mundo)*, 33, 34, 40, 46*f*, 51, 90, 143–149; interview, 143–49; locations, 57; protagonists, 57; Raging Sun, Raging Sky *(Rabioso sol, rabioso cielo)*, 14, 33, 35, 52, 72, 144, 145,

Hernández, Julián, (cont'd)
147; *Rubato lamentoso*, 146; *Selección*, 54, 56; short films, 40, 52; style of, 57–59; success in Mexican films, 148; themes, 50; A Thousand Clouds of Peace Encircle the Sky, Love, Your Being Love Will Never End *(Mil nubes de paz cercan el cielo, amor, jamás acabarás de ser amor)*, 90, 143–145; Young Man at the Bar Masturbating with Rage and Nerve *(Muchacho en la barra se masturba con rabia y osadía)*, 56, 90
Hernández Cordón, Julio, 137–140
heterosexual, 38
Historic Center, 3
Hitchcock, Alfred, 87, 90
HIV, 4
Hollywood Reporter, 35, 50, 98, 104
Holmlund, Chris, 62, 63, 70
hombres heterosexuales, 38
hombres/hombres normales, 37, 38
Homocinefilus (blog), 67
homoeroticism, 14, 36
homophobia, 11, 15, 16, 46, 88, 98, 99
homosexuales/gays, 38
homosexuality and Mexican television (incompatibility of), 108
Homosexual Liberation Front (Mexico City), 4
Homos magazine, 23, 24
The Hotel of Secrets *(El hotel de los secretos)*, 140
hybridity, 7

I Am Happiness on Earth *(Yo soy la felicidad de este mundo)*, 33, 34, 40, 46f, 51, 58, 90, 143–149; synopsis, 46, 47
Ibarra, Epigmenio, 112, 114, 129, 140
Ibarra, Eréndira, 138
Iberoamerican Federation of Cinematic and Audiovisual Produces, 34
identity politics, 13; definition of terms, 37, 38; *muxe* identity, 102; transgender, 62 (*See also* transgender documentaries)
India, LGBT festivals (Kashish), 12, 13
Infante, Pedro, 39
El Informador, 67
Institutional Revolutionary Party (PRI), 4
internacionales, 37, 43
Internet: free aggregation sites, 8; webseries, 8 (*See also* webseries)
interviews (Julián Hernández), 143–49
I Promise You Anarchy *(Te prometo anarquía)*, 137–140

Jara, Rubén, 109, 111
La Jetée, 9
Joa Silar, Bilai, 123
Johnson, Martha, 64
Jones, Rebecca, 114
La Jornada, 15, 52, 121
José, José, 33, 140

Kahlo, Frida, 14
Kashish (India festival), 12, 13
Kluge, Alexander, 144

labels, 13
La Bruce, Bruce, 15
Ladies Dress Designer *(Modisto de señoras)*, 140
Ladyboys! *(Mujercitos!)*, 64
Latin American Tower, 3
Lee, Ang, 35
lesbians, 48, 108, 127, 131; At the End of the Rainbow *(Al final del arcoíris)*, 20; Everyone Has Someone But Me *(Todo el mundo tiene a alguien menos yo)*, 85, 86, 89f, 93, 94, 106; films, 31; identities, 62; themes, 9, 16
Letras Libres (literary monthly), 67, 68
LGBT (lesbian, gay, bisexual, transgender): festivals, India (Kashish), 12, 13; media, 1; shift in attitudes toward, 5
Life in the Mirror *(La vida en el espejo)*, 108, 109, 113f, 114–120, 122; synopsis, 113
locations: Hernández, Julián, 57; Mexican, 144
López, Aida, 134
López, Sergio Raúl, 26, 27
López Pumarejo, Tomás, 22, 23
Los Angeles, California, 13
Luna, Diego, 115

machismo, 38
Made in Bangkok, 77f, 79, 81, 82, 83
mainstream movies, 85–106; Everyone Has Someone But Me *(Todo el mundo tiene a alguien menos yo)*, 85, 86, 89f, 91–94, 104, 106; Four Moons *(Cuatro lunas)*, 86, 94f, 95–100, 104–106; Tropical Lipstick *(Carmín tropical)*, 86, 101f, 102–104, 106
Maitra, Ani, 12, 13
major feature films. *See* films
Make Yourself at Home *(Faites comme chez vous)*, 107
maricones, 46
Marker, Chris, 9

164 - Index

masculine: privilege, 39; self images, 69
masturbation, 47, 56
McQueen, Alexander, 42
Mecos, 7, 8, 24, 25, 30, 41, 87
media in Mexico, 1
Medina, Cuauhtémoc, 5
Méndez, Raúl, 122, 124
Mexican films. *See also* films: box office, 35; success in, 148
Mexican Human Rights Film Festival, 78
Mexican locations, 144
Mexican Men, 2, 140
Mexican Zapatistas, 88
MeXiCoCitizens *(DFÑOS)*, 23
Mexico: culture, 37; media in, 1; social change in, 85, 86, 99
Mexico City, Mexico, 3, 93, 94; Homosexual Liberation Front, 4; Mix Festival (2014), 7
Mexico City Ministry of Culture, 79
Mexico Is Spelled with a J *(México se escribe con J* [Schuessler and Capistrán]), 5
Midaq Alley *(El callejón de los milagros)*, 97
middle class, increase in size of, 5
Milenio, 72, 73, 108, 122
Mil nubes de paz cercan el cielo, amor, jamás acabarás de ser amor (A Thousand Clouds of Peace Encircle the Sky, Love, Your Being Love Will Never End), 90, 143–145
Mirada de mujer (A Woman's Look), 110, 111
Miss Russia, 80, 81. See also *Made in Bangkok*
Mix Festivals, 7, 8, 9, 10*f*, 11, 13, 30, 34, 87; advances for gays, 16; audiences, 18; global locations of, 14; history of, 17; homophobia, 15, 16
Modisto de señoras (Ladies Dress Designer), 140
Monsiváis, Carlos, 3
Morelia International Film Festival, 2, 35, 78, 104, 149
Morganna, 78, 79, 80, 84. See also *Made in Bangkok*
Morir de pie (To Die Standing Up), 66*f*, 67–70, 72, 79
movies. *See* films; mainstream movies
Muchacho en la barra se masturba con rabia y osadía (Young Man at the Bar Masturbating with Rage and Nerve), 56, 90
Mujercitos! (Ladyboys!), 64
muxe identity, 102

narration: Disrupted *(Quebranto)*, 74; voice-of-God, 67, 70

National Fund for Culture and the Arts *(Fondo Nacional para la Cultura y las Artes)*, 13
nationalism, 86
Neighbors *(Vecinos)*, 107, 108
Netflix, 104
networks, Azteca, 2
The New Homosexual Film Festivals, 34
New Queer Cinema (Rich), 2
New Queer Cinema (United States), 144
Nichols, Bill, 63
The Night Is Young: Sexuality in Mexico in the Time of AIDS (Carrillo), 36
Noir magazine, 79
No One Can Live Here *(Aquí no hay quien viva)*, 107
Northless *(Norteado)*, 101, 104

Only with Your Partner *(Solo con tu pareja)*, 3, 4
Orozco, Guillermo, 110, 111
Ortiz Urquidi, Moisés, 138
Las oscuras primaveras (The Dark Springs), 111
O'Toole, Gavin, 4
Our Lady of Guadalupe, cult of, 39
Ozon, François, 9

panels, festivals, 12
pasivos, 37, 43
Peña Nieto, Enrique, 121
Pereda, Nicolás, 48
Perezcano, Rigoberto: Tropical carmine *(Carmín tropical)*, 86; Tropical Lipstick *(Carmín tropical)*, 101*f*, 102–104, 106
Pérez-Sánchez, Gemma, 64
Pinoles, Doña, 74
Piñón, Juan, 22
Plato, 43
Podalsky, Laura, 36, 44–46
politics, identity, 13
PornHub, 8
pornography, 1, 25. *See also* films
Porn Studies academic quarterly, 26
Portal, Andrea, 93
posthegemony, 88, 99, 104
Preciado, Paul B., 64
PRI (Institutional Revolutionary Party), 15, 85
prostitution, 42, 96
protagonists (Julián Hernández), 57
La putiza (The Beating), 27

Quebranto (Disrupted), 70*f*, 73–77, 79, 83
queer characters, 3
Queer Masculinities in Latin American Cinema (Subero), 41

Raging Sun, Raging Sky *(Rabioso sol, rabioso cielo)*, 14, 33, 35, 52, 72, 144, 145, 147
Ramírez, Alan, 34, 51
Realities and Desires (*Realidades y deseos* [Casarin]), 65
Rear Window, 87, 90
Rebelde (Rebel), 19, 20
The Reinvention of Mexico, 4
Reygadas, Carlos, 4, 5, 44, 105
Richards, Stuart, 12
Ripstein, Arturo, 5
Ripstein, Gabriel, 71
Rod-García, Jero, 90
Rodríguez, Gabino, 48
Rodríguez, Joaquín, 71
Romero, Bernardo, 114
Rubato lamentoso, 146
Ruiz, Stefan, 139
rules of art (Bourdieu), 58
Russo, Vito, 14

safe sex, 37
Sánchez, Diana, 12
Sánchez, Jorge, 34
Sánchez Prado, Ignacio, 4
Schulz-Cruz, Bernard, 5
Selección, 54, 56
Serrano, Antonio, 114
sex, explicit in films, 50
Sex, Shame, and Tears *(Sexo, amor y lágrimas)*, 114
Sexo, amor y lágrimas (Sex, Shame, and Tears), 114
El sexo débil (The Weaker Sex), 120*f*, 123–127, 137; synopsis, 121
sexual diversity, 3, 12
sexual diversity, themes, 9
sexual identities, 37, 38
sexuality, fluidity of, 13
sex workers, 82, 95, 96. *See also* prostitution
SeXXXcuestro (XXX Kidnap), 26, 28, 29, 30
short films, 1. *See also* films; *Atmósfera*, 55; *Bramadero*, 28, 52, 53, 54, 56; *David*, 49; Hernández, Julián, 40, 52; *Mexican Men*, 2; *Rubato lamentoso*, 146; *Selección*, 54, 56; A

Summer Dress *(Une Robe d'été)*, 9; Young Man at the Bar Masturbating with Rage and Nerve *(Muchacho en la barra se masturba con rabia y osadía)*, 56
Silva Baena, Armando, 24, 26, 30
Simplemente Maria, 134
Slim, Carlos, 4
social change in Mexico, 85, 86, 99
social media, 31
Society of Cinema and Media Studies (SCMS), 12, 26, 87
SoDoMe bathhouse (Mexico City, Mexico), 15, 93
Solo con tu pareja (Only with Your Partner), 3, 4
States News Service, 36
stereotypes, 5
Straayer, Chris, 63
street vendors *(ambulantes)*, 76
studios, Mecos, 30. *See also* Mecos
subcultures, gay, 42
Subero, Gustavo, 36, 41, 42, 46, 51, 58
Supernatural, 87

Talk to Her *(Hable con ella)*, 48
Teena, Brandon, 63
Telemundo, 123
telenovelas, 22, 95; Argos, 107–35 (*See also* Argos, telenovelas); critics, 109; format of, 110; homosexualization of, 23; The Hotel of Secrets *(El hotel de los secretos)*, 140
Telenovelas en México: Nuestras íntimas extrañas (Telenovelas in Mexico: Our Intimate Strangers), 109
Televisa, 123, 131, 134
television, 1. *See also* telenovelas; *Capadocia*, 100; *Guau*, 61; Neighbors *(Vecinos)*, 107, 108; *Rebelde* (Rebel), 19, 20; *Supernatural*, 87
Te prometo anarquía (I Promise You Anarchy), 137–140
Thailand, 2
The Dark Springs *(Las oscuras primaveras)*, 111
themes: Hernández, Julián, 50; lesbian, 9; lesbians, 16; sexual diversity, 9; visual fixation, 11
Time Out México magazine, 17
To Die Standing Up *(Morir de pie)*, 66*f*, 67–70, 72, 79
Todo el mundo tiene a alguien menos yo (Everyone Has Someone But Me), 85, 86, 89*f*, 91–94, 104, 105, 106; synopsis, 89

tolerance, 3
Tovar, Luis, 73
Tovar Veladre, Dergio, 104, 105
The Transformation of Film into Music *(La transformación del cine en música)*, 145
transgender documentaries, 61–84; To Die Standing Up *(Morir de pie)*, 66f, 67–70, 79, 80; Disrupted *(Quebranto)*, 70f, 72–80, 83; Made in Bangkok, 77f, 79–83
"Transgender Mirrors: Queering Sexual Difference" (Straayer), 63
transmedia, 7, 88
"Transnational Conversations in Migration, Queer, and Transgender Studies: Multimedia Storyspaces" (Pérez-Sánchez), 64
Tres Tercios, 7, 30; At the End of the Rainbow *(Al final del arcoíris)*, 19–24. 30; webseries, 29
Treviño, Marco, 121
Tropical Lipstick *(Carmín tropical)*, 86, 101f, 102–104, 106; synopsis, 101
Truffaut, François, 86
Trujillo, Michelle, 61, 62
TVNotas magazine, 5, 100,
TVyNovelas magazine, 5, 23, 61
Twitter, 31, 99

UNAM (National Autonomous University), 41, 145
Une Robe d'été (A Summer Dress), 9
United States, 5; documentaries, 63; gay characters on television, 108; New Queer Cinema, 144
El Universal, 78
Univision, 123

van Hoeij, Boyd, 98, 104
Vargas, Diana, 12
Vargas, Susana, 64, 65
Variety, 35, 36
Vecinos (Neighbors), 107, 108

Velarde, Sergio Tovar, 95; Four Moons *(Cuatro lunas)*, 86, 94f, 96–100, 106
La Verga Parada (The Erect Cock), 27
Veronika Voss, 48, 144
Vestido, 9, 18, 34
La vida en el espejo (Life in the Mirror), 108, 109, 113f, 114–120, 122; synopsis, 113
Villela, Arturo, 44, 48
Vimeo, 9, 45
visual fixation (theme), 11
Voces en Tinta (Voices in Ink), 64
voice-of-God narration, 67, 70
von Sternerfels, Emilio, 49

The Weaker Sex *(El sexo débil)*, 120f, 123–127, 137; synopsis, 121
webnovelas, 1, 22
webseries, 7, 8; At the End of the Rainbow *(Al final del arcoíris)*, 19–24, 30, 44; Tres Tercios, 29
websites, 9
Williams, Linda, 26
Williams, Raymond, 110
women, transgender documentaries, 62. *See also* transgender documentaries

XTube, 8
XVideos, 54
XXX Kidnap *(SeXXXcuestro)*, 26, 28, 29, 30

Yazpik, José María, 108, 113–115, 133
Yo soy la felicidad de este mundo (I Am Happiness on Earth), 33, 34, 40, 46f, 51, 58, 90, 143–149; synopsis, 46, 47
Young Man at the Bar Masturbating with Rage and Nerve *(Muchacho en la barra se masturba con rabia y osadía)*, 56, 90
YouTube, 9, 31, 126

Zanzibar International Film Festival, 78
Zielinski, Ger, 12
Zona Rosa, 3

www.ingramcontent.com/pod-product-compliance
Lightning Source LLC
Chambersburg PA
CBHW052052220426
43663CB00012B/2533